KEVIN J. HARTY
JOHN KEENAN

LA SALLE UNIVERSITY

Writing for Business and Industry
Process and Product

MACMILLAN PUBLISHING COMPANY
NEW YORK

For Jamie and Alice

Macmillan Publishing Company
866 Third Avenue, New York, New York 10022

Collier Macmillan Canada, Inc.

LIBRARY OF CONGRESS CATALOGING-IN-PUBLICATION DATA

Harty, Kevin J.
  Writing for business and industry.

  Includes index.
  1. Business report writing.   2. Technical writing.
I. Keenan, John.   II. Title.
HF5719.H36 1987   808′.066651   86–2804
ISBN 0–02–351400–0

PRINTING: 1 2 3 4 5 6 7   YEAR: 7 8 9 0 1 2 3

ISBN 0-02-351400-0

Preface

Ability to express ideas cogently and ideas persuasively—in plain English—is the most important skill to leadership. I know of no greater obstacle to the progress of good ideas and good people than the inability to compose a plain English sentence.   —JOHN D. DEBUTTS, FORMER CHAIRMAN, AT&T

Writing for Business and Industry: Process and Product is a comprehensive college-level textbook for students in courses in business, technical, and professional writing. Business and technical people already on the job will find *Writing for Business and Industry* an equally comprehensive reference book.

As our subtitle suggests, we are concerned with both process and product. Readers will find in the chapters that follow detailed discussions of resumes and other job-getting documents, letters, memos, and a variety of reports. First, however, readers will find an even more detailed application of the writing process to the needs of business and industry.

Our approach to the writing process—a systematic analysis of the different situations in which business, technical, and professional writers find themselves daily—is based on a consideration of purpose, audience, format, evidence, and organization. Throughout this book, we use the acronym PAFEO in discussing our five-part approach to the writing process.

Whether discussing process or product, we have based what we have to say on documents culled from our work as consultants or supplied by our associates and former students in business and industry. These documents come from the files of banks, insurance companies, various businesses and industries, computer firms, newspapers, law firms, hospitals, nonprofit organizations, government agencies, and academic institutions. We have purposefully structured our textbook this way to provide readers with real examples from the world of work rather than with generic samples tailored to meet imaginary or simulated case studies.

Writing for Business and Industry is divided into three parts and an appendix. In Part I, we first introduce readers to the idea of writing as process. We then discuss PAFEO, the different parts of this process—Purpose, Audience, Format, Evidence, and Organization—in greater detail. We conclude Part I with an application of the principles of revision to the writing process as it applies to the world of work.

In Part II, we discuss a variety of writing products that readers write or can expect to write on the job. At times, where we find them helpful or appropriate, we provide checklists for readers to follow in their own writing. At other times, we introduce a variety of problem-solving strategies for readers to adapt to their own writing situations. Throughout, our aim is to be flexible and not dogmatic. There is no one way to write a resume, a letter, a memo, or a report. Writers in business and industry need a variety of strategies and checklists to rely on as they face their daily writing challenges.

We begin Part II with a detailed discussion of writing products of most immediate concern to our readers who are still in the classroom. These products include resumes and cover letters. (Readers already on the job but contemplating a job or career change will also find valuable information in this chapter.) We move from a discussion of resumes and cover letters to a series of chapters on business letters, memos, and reports. Our goal here, as elsewhere, is to be as comprehensive and as practical as possible. Although Part II is specifically entitled "Products," we continue to discuss the writing process as it applies to the various products that we examine, because, in our view, process and product are inseparable.

In Part III, we briefly discuss a number of mechanical and grammatical tools that busy writers need to have at hand. Our approach here is selective, based on common sense and our experience as teachers and consultants. We introduce the grammar that busy writers in business and industry need to know.

A brief appendix discusses the application of computers and word processors to the writing process and to the writing products of the world of work. Although computers and word-processing software will never fully take the place of the hard work that good writing requires, they do offer innovative approaches to revising and editing to which we wish to introduce our readers.

A complete index and a detailed table of contents provide readers with easy access to the materials we discuss in this textbook. Inside the front and back of the book's cover we provide a PAFEO checklist for quick reference.

Throughout *Writing for Business and Industry,* we follow a common organizational pattern. Each chapter begins with a brief discussion and moves to an analysis of a writing sample from the world of work. The balance of the chapter alternates discussions of key points with analyses of their applications in successive writing samples. Each chapter is followed by a set of exercises. The first exercises are designed for classroom discussion, although they can be adapted easily to writing assignments. The latter exercises ask readers to write or rewrite documents appropriate to the world of work.

A special feature of *Writing for Business and Industry* is that it is a writing textbook that shows readers writing samples and asks them to write work-related materials from the very beginning. Our bias, we happily confess, is toward the practical rather than the theoretical.

Acknowledgments

In writing any textbook, authors incur a great many debts, as many as possible of which we are pleased to acknowledge here. Our students—both at La Salle University and in corporate writing programs for Philadelphia Life Insurance Company, the Federal Reserve Bank, M. A. Bruder and Sons, Philadelphia Newspapers Incorporated, First Pennsylvania Bank, the Board of Pensions of the United Presbyterian Church USA, and Computer Partners, Inc.—have been willing and helpful critics.

Our five-part PAFEO approach to the writing process was first developed by John Keenan for his earlier project, *Feel Free to Write* (New York: Wiley, 1982). Our footnotes show our debts to other writing instructors and their textbooks.

Macmillan commissioned reviews of our book in its various stages of manuscript development. For their careful readings and thoughtful comments, we are grateful to R. Neil Dortch, University of Wisconsin-Whitewater; Mary Alice Griffin, Valdosta State College; John S. Harris, Brigham Young University; Judith R. Levine, University of Kentucky; Richard Profozich, Prince George's Community College; Jack Selzer, The Pennsylvania State University; and Aileen Chris Shafer, West Virginia University.

Our own colleagues at La Salle University have been equally generous with their time and advice. Through its Leaves and Grants Committee, headed by Brother Emery Mollenhauer, F.S.C., La Salle demonstrated faith in our project by awarding us each Summer Research Grants. Brother James Muldoon, F.S.C., Dean of Arts and Sciences, was also supportive of our efforts, as were our colleagues in the English Department: Margot Soven, Jim Butler, John Kleis, Richard Lautz, Jack Sweeder, and Claude Koch; and our colleagues at the University Library: Joy Collins, Don Root, and Ellen Wall. P. Douglas McCann, our colleague in the Finance Department, and his student, Joseph E. Klink, generously helped with the sample computer graphics. Connie Callan and Francine Lottier applied their typing and word-processing skills—along with considerable patience and abundant humor—to various drafts of the manuscript. For his assistance in the compilation of the index, we are grateful to Martin J. Piccoli.

For additional support in a variety of ways and under various circumstances, we are grateful to Susan Anker, John J. Harty, Maureen Harty, and the staff of the College Division at Macmillan, especially Eben Ludlow and Wendy Polhemus. Thanks finally to Jamie and to Alice and the family; they endured.

K. J. H.
J. K.

Contents

CHAPTER 7   **Messages That Inform**   **147**

PART I

PROCESS

CHAPTER 1

The Writing Process

Let's begin this book by recognizing a fundamental truth. Life holds few certainties. You can count on having to cope with taxes, change, and death. And you can probably count on doing some writing as part of your job. (The latter is not quite as certain as death and taxes and change, but it's a probability any gambler would regard as a sure thing!)

If you are now employed, you may have recognized this fact already; it may be part of the reason you're back in the classroom. If you are a full-time student, you already know that writing plays an important role in your education. If anything, that importance will grow when you enter the working population. Virtually every career a college graduate might choose requires writing. You will write application letters and resumes to get a job, and once employed you will discover that the amount of writing you do increases with every promotion. In fact, your writing skills may directly influence whether you get that promotion.

This course, then, is an important part of your education, and this book may be an important help to your career. If you work on the principles and techniques presented in these pages, you will develop writing habits that free you to communicate your ideas clearly and confidently.

The Three Stages of the Writing Process

For many years, the process that went into creating that final draft was thought to be too individualized to be taught. As a result of research into the writing habits of effective writers, we now know better. We realize that good writers generally follow a *process* when they write, and that the process consists of three stages:

1. *Planning,* which includes developing and selecting content, clarifying the purpose, and analyzing the needs of the audience.
2. *Arranging* the content in a way consistent with the writing situation, which includes the message, the audience, and the purpose.
3. *Revising* to ensure that the writing achieves its purpose for the intended audience by using the most effective arrangement and the appropriate style.

While effective writers may vary in carrying out these stages, they do not skip any of them when tackling a serious and challenging writing task. (Admittedly, writers may not revise their personal letters or arrange the content of notes to their children, but they recognize the need to revise in their business writing.) Ineffective writers, on the other hand, frequently give little attention to stages 1 and 3, and they interpret stage 2 mostly in terms of content, ignoring matters of purpose and audience. We argue that failure to carry out the complete writing process contributes significantly to the weaknesses of the products they produce.

The Results of Overemphasizing Product

How about you? Do you make things difficult for yourself by trying to arrive at a Perfect Product in one inspired burst? Do you attack each writing job by just starting to write, hoping that momentum will carry you along to a successful completion? This method can work—once in a while. (Even weak-hitting pitchers hit a home run once in a while.)

Or do you agonize over each sentence, trying version after version, substituting one word or phrase for another until you're confused and exhausted?

Or do you think that writing depends on inspiration, on waiting for the magic moment when the solution to your writing will appear like the yellow brick road and lead you to your goal—the Emerald City, a shimmering perfection that all will admire?

However common and popular these approaches may be, they share a common misconception. Because they focus too much on the finished piece of writing—the product—they ignore the experience and research that show better ways of solving the many complex problems associated with effective writing.

The Advantages of the Process Approach

Perhaps the greatest advantage of focusing on the writing process rather than on the finished product is that you are free from the pressure of trying to do everything at once. You don't have to make the leap from blank page to Perfect Product. Once you recognize that writing consists of many smaller choices rather than one momentous leap, your whole attitude toward writing can change. You can take control. You can recognize that writing involves choices, decisions about the best way to achieve a series of goals.

When you stop thinking of writing as only a means to an end, you will be free to enjoy the writing process as problem-solving and discovery. Your successes will not depend only on the perfect (and therefore impossible) final product but on discovering your meaning and your method through systematic thinking about the decisions you face. These decisions include:

1. Determining the exact purpose your writing aims to achieve.
2. Focusing on the intended audience and thinking about that audience's needs and attitudes toward the message and the writer.
3. Choosing a format best suited to the purpose and audience.
4. Gathering the information and examples needed to explain and support the purpose so the audience can understand and accept it.
5. Arranging the content in an order that will make the purpose clear to the intended audience.

Writing is not, however, as simple as 1-2-3, and it cannot be accomplished in a step-by-step arrangement. The five decisions we just outlined are interrelated, and good writers are constantly moving back and forth freely among the stages of the writing process as they solve some problems and discover new ones. Therefore, any strategies we describe in this book must be viewed as flexible; some writing jobs will require more of one strategy than another; some strategies won't work well in a given writing situation and will have to be replaced with another approach.

An analogy to sports may help explain what we mean. A football coach prepares a strategy for a game: "We will establish the running game until the opposition is vulnerable to the passing game." He has a purpose which he thinks will be effective, given what he knows about his audience (the opposition). But as the game progresses, the running game doesn't. The coach must adjust his original plan at halftime and try another strategy. If he is a good coach, he has many others he can try, based on the needs of his team and the reactions of his opponents.

And so it is with you as a writer. You too must be flexible in your systems. If one systematic approach isn't working as well as you'd like, it's important to have another to try.

In this book, we'll introduce you to several of the techniques writers have developed to solve their problems. We will emphasize one particular approach which we ourselves have found sound and useful. Both our students and business people with whom we have worked seem to agree.

The PAFEO Process

Groucho Marx made "the secret word" famous on his 1950s television show, *You Bet Your Life.* If a contestant said the secret word in the course of bantering with Groucho, a toy duck descended from the ceiling, everyone cheered, and the surprised contestant won a cash bonus.

We have a secret word of our own, and we're going to share it with you throughout this book and show you how it works in various business writing situations. While you won't get money for saying it, as Groucho's contestants did, there is a good chance it will help your earning power in the future by making you a better writer.

The word is PAFEO. Actually, it's not a word at all. It's an acronym that functions as a mnemonic, a memory aid to remind you of what you have to consider as you write. It works throughout the three stages of the writing process—the planning, arranging, and revising. Whether you have to write a letter, a memo, a proposal, or a report, you will find the PAFEO process an invaluable guide to what's important in writing. Here's what each letter of the acronym means.

P is for Purpose
A is for Audience
F is for Format
E is for Evidence
O is for Organization

By the time you've finished this course, we hope you will have become so accustomed to using the PAFEO process that you will not even be conscious that you are using it. In the meantime, however, you are going to see the acronym frequently as we show you how it can work in various situations.

PAFEO: An Overview

At this point, it may be helpful to take a brief look at each of the five points of the process. Although we must discuss the items one at a time, we remind you that the five points are not step-by-step in arrangement; you cannot check off one at a time, using the order in which they appear in the acronym. They describe a dynamic process, not a series of mechanical steps. For example, you cannot decide *purpose* without considering the needs of your *audience*. *Purpose* and *audience,* in turn, help to determine the kinds of supporting *evidence* you will need, the *organization* you will select, and the *format* you will design.

PAFEO is not a magic word intended to fool you into thinking that the complex task of communicating in writing is easily solved with a simplistic gimmick. We see it instead as a way of remembering a sound, basic strategy for thinking through a problem in writing.

Purpose

Good business writing aims at achieving a particular goal with a particular audience. Purpose and audience interact, each affecting the other. Having a vague goal that does not consider the problems of the audience is not the same as having a real purpose for writing.

To clarify your purpose, try to put into words exactly what you want the reader to think or do as a result of reading what you have written. Write out your *statement of purpose* in a complete sentence, keeping it as concise as you can. By being as specific as possible, you will help yourself by defining the problem you have to think about. Focusing on the desired goal with a particular audience in mind will help you to understand the problem your writing should aim to solve.

When we started this book, we could have written a statement of purpose this way:

> We want to write a textbook for students of business writing.

This sentence is too general though. It doesn't spell out what we want the reader to be able to think or do. So let's try again.

> We want to write a textbook for students of business writing so that they will learn to write more effectively by using *The PAFEO Process* to solve their job-related writing problems.

Isn't that a considerable improvement? Notice that we have forced ourselves to identify the intended audience more precisely. We now have a goal: we know what we want our audience to be able to do as a result of reading our book, and we have an idea of how we are going to achieve that goal—by showing readers how PAFEO can help them with job-related writing. We now have a better conception of what problem we are going to try to solve for the reader.

Although we show you here only two versions of our purpose statement, honesty compels us to admit that we labored and fought our way through at least a half dozen other versions, each getting painstakingly closer to what we wanted. But because we knew what we wanted in order to create an effective purpose statement, we were able to make choices of wording much more easily and effectively. Focusing on the purpose and defining the problem to be solved will make many of your decisions easier.

Audience

On the same day, the editors of the three major New York daily newspapers decided that three different stories deserved their readers' immediate attention.

The *Post* ran the following headline:

170 BUSTED IN BIG POT SWOOP

The *Daily News* ran a two-part headline:

NEWSBOY-SLAYING TEEN:
THE DEVIL MADE ME DO IT!

The *Times,* which doesn't use a banner headline, centered a picture on page 1 of the President conferring with the Secretary of State. Under the picture, there was an article with this headline:

SCHULTZ SAYS HE IS CONFIDENT SYRIA
WILL CONSENT TO LEBANON PULLOUT

The news was the same for all three newspapers that day. Why the difference in headlines? Why the three different decisions about that day's "hot" news story?

The difference is in audience. The *Post* and the *Daily News* appeal to different audiences than the *Times* does. Readers of these three newspapers look to the newspaper of their choice for different kinds of information. The difference in readership is not based solely on class distinction or snob appeal. Rather, that difference is based on what each newspaper does for its readers, on how well the newspaper meets the needs of its readers.

Readers wanting news at a glance don't turn to the *Times.* Readers packed into city buses and subway cars prefer the *Post* and the *Daily News,* whose tabloid format is designed for easy reading while standing. The many pictures with paragraph-long captions and the generally short length of the stories are also designed to make these papers easy to read. On the other hand, readers need elbow room to tackle the *Times,* whose stories from page 1 may be continued on pages 54 or 79 in a second, third, or fourth section of the newspaper.

Time Inc. has successfully staked out three different audiences and met their needs with magazines as different as *Time, People,* and *Sports Illustrated.* Each provides readers with the information and entertainment they want in the form they want. Should any one of these magazines fail to meet the needs of the audience it aims at, it will soon be out of business.

Obviously, you say. No one can afford to ignore the reader. It's true that no one starts out intentionally ignoring the reader, but it is just as true that many writers end up having done just that. Why? They simply get wrapped up in the difficulties of trying to say what they mean. In the struggle to get ideas into words, they settle for writing that works well enough for them without asking whether it also works well for a reader. The writing they produce is, therefore, *writer-based* rather than *reader-based.*[1]

---

[1] We borrow the distinction between reader-based and writer-based prose from Linda Flower, whose influential textbook—*Problem-Solving Strategies for Writing,* 2nd ed. (San Diego: Harcourt Brace Jovanovich, 1985)—has informed much of our own thinking about the writing process.

If you want to write reader-based prose, you must practice analyzing your purpose and your audience until you have made such analysis a habit of thinking. Communication in business writing depends heavily for its success on how well you meet your audience's needs. In writing papers for your instructors in high school or college courses, you may not have devoted much time or effort to audience analysis. You wrote for the professor, a captive audience committed to reading all papers. But the writing you do in business offers no such guarantees. If you're dull and disorganized, you stand a good chance of losing your reader quickly. However important your message is, it will not get through. If you want it to get through, you will have to take care to make it reader-based. Who will be reading this message? If there will be different audiences, such as managerial and technical people, what do they need to know? What's the best way of getting the information they need across to them? What tone should you take in order to ensure they will read this message without setting up roadblocks along your paths of communication? These are the kinds of questions you will learn to ask yourself. Granted, each reader is an individual with individual needs; nevertheless, you may count on these general characteristics in virtually all readers of business communications. Remember to keep these characteristics in mind from the start.

THREE CHARACTERISTICS YOUR READERS SHARE 1. *All readers are busy.* You can never assume that your readers will want to read your letter, memo, or report. They have a hundred other things that ought to be done immediately. Your message is part of that endless pile known ingloriously as "paperwork." You can't waste your readers' time. You have to persuade them quickly that the document you present is worth taking time to read.

Try this simple test on the next document you write. Can you persuade your reader in the first ten seconds of reading that the document is worth completing or routing to the appropriate person? If not, such failure is much worse than a low grade on a classroom assignment. Not only have you missed an opportunity to communicate your message, but you have also made a poor impression on your reader that may carry over to future dealings. So it's important that you learn how to get a busy reader's attention and hold it.

In *Confessions of an Advertising Man,* David Ogilvy, one of the leading figures in advertising, related the following anecdote:

> Vic Schwab tells the story of Max Hart (of Hart, Schaffner, Marx) and his advertising manager, George L. Dryer, arguing about long copy. Dryer said, "I'll bet you ten dollars I can write a newspaper page of solid type and you'd read every word of it."
>
> Hart scoffed at the idea. "I don't have to write a line of it to prove my point," Dryer replied. "I'll tell you the headline: THIS PAGE IS ALL ABOUT MAX HART."[2]

[2] David Ogilvy, *Confessions of an Advertising Man* (New York: Atheneum, 1963), p. 110.

Advertising copy writers know how to catch their readers' attention immediately. You can do the same thing in your letters, memos, and reports by

- not wasting your reader's time
- coming to your main point as soon as possible
- being specific about your purpose in writing

To demonstrate the importance of these three points, we would like you to examine the following two versions of a memo being sent to all soon-to-retire employees by the Vice-President for Benefits and Compensation. Neither memo is presented as a horrible example, but one is more effective than the other because it carries out the three suggestions above. As a busy employee, which version would you be more inclined to read and act upon immediately? Why?

```
August 18, 1986

TO:    William D. Viglione
       Accounting, C-346
FROM:  Amanda Wing
       VP, Benefits & Compensation


Perhaps you're the kind of person who has never given more
than a fleeting moment of serious thought to the idea of
retirement.

We have found that retirement must be planned to be
successful. We have spoken with many retirees and have
been doing extensive research on the topic of retirement.
We have been concerned that our retirees deal with and
effectively manage the varied changes in lifestyle,
financial obligations and resources, and health that
retirement can bring.

Just as you planned for every other milestone in your
life, you should plan for your retirement, whether you
have been with your employer 35 or many fewer years. Now
is the time for such planning. You must keep in mind that
retirement can last a quarter of your life. You should,
therefore, prepare for it as carefully as you prepared for
your first career.

We have developed a Retirement Planning Seminar which will
deal with the various concerns that people have as they
approach this important stage in life.

The seminar will meet on September 12 and then every
Monday thereafter through October 24. The seminar will
meet in the multipurpose room in Building A. Each meeting
```

will run from 9:00 am until noon. Your manager has been advised of these meetings in order to provide you with time off to attend them.

At this time, we wish to extend to you and your wife a hearty invitation to attend these seminar meetings.

To help in planning these seminar meetings, we need to know whether you will attend alone or with your wife. Please let me know by August 23.

We will try to provide you with useful information, and we look forward to your joining us.

AW/kk

August 18, 1986

TO: William D. Viglione
          Accounting, C-346
FROM: Amanda Wing
          VP, Benefits & Compensation
SUBJECT: Retirement Planning Seminar

Benefits and Compensation wants to help you prepare for your retirement by inviting both you and your wife to a seminar on retirement planning.

The seminar will meet weekly from September 12 through October 24, Monday from 9:00 am until noon. All meetings will be held in the multipurpose room in Building A. Your manager knows you will be attending the seminar meetings.

Please let me know by August 23 whether your will attend alone or with your wife.

I look forward to your joining the seminar.

AW/kk

2. *All readers expect writing to have a purpose and an arrangement that the writer will make clear.* Your reader can't be expected to plow through sentence after sentence attempting to figure out why you're writing. Nor will your reader appreciate your shifting gears in the middle of what you write. What starts out as "For Your Information" cannot suddenly turn into "Do this or else." The reader ought to be able to see the goal your writing is headed towards and distinguish between your major and minor ideas.

You can help meet these expectations by providing both internal and external signals that guide the reader along the lines of your thinking.

What is the purpose of the following memo? Does the arrangement help the reader see that purpose quickly and clearly?

April 13, 1986

TO: Rita Harris
        Accounts Payable
FROM: Montgomery Kaplan
        Claims Investigation
RE: William York
        Acct. # 9-0876-YO-89-9865543

Mr. York contacted his service representative about a nonrecorded payment of $32.20. The service representative contacted me sometime in early February.

The service representative requested duplicate copies of all checks Mr. York sent us in the last six months. The copies were received 2/10/86. The service representative sent these copies along with an explanation to your department on 2/12/86.

On 2/21/86, your office requested a second set of copies of the checks. This second set arrived on 3/12/86 and was sent to your department immediately along with a further note of explanation.

On 3/18/86, our president received a formal complaint on Mr. York's behalf from the Better Business Bureau. He immediately contacted me.

Upon investigation, I discovered that Mr. York's check was wrongly credited to another account.

Please send Mr. York a receipt for his payment. The attached copy of the payment record will supply you with the information you need.

This memo comes from the files of a large insurance company, and it's a toss-up whether we should feel sorrier for Rita Harris or Mr. York. All Harris needs to know is contained in the last paragraph. The memo illustrates reader-based writing: Kaplan is rehashing the notes in his file. But Harris doesn't need these notes. She simply needs to know that she is to issue a receipt to satisfy an already justifiably angry Mr. York. Kaplan has not remembered how busy his reader is. He has wasted her time with a chronological narrative that appears to be simply informative background, and he has waited until the end of the memo to get to his real purpose. If Kaplan thought Harris needed the background, he should have given it after making his primary purpose clear immediately. As the memo is now written, Harris may decide she doesn't need all the information, stop reading, and never get to the real purpose. And poor Mr. York will be left waiting even longer for his receipt.

3. *All readers know less about the subject of this communication than you do.* This third characteristic places an added burden on any writer. As Steven Pauley pointed out:

> In the world of business and industry, you will generally write for an uninformed reader. If that seems odd, think of it this way: If the reader knew more about the subject than you did, he or she would be the writer and you would be the reader.[3]

As a writer, you are presumably writing because you may have something important to say to your reader. As a writer, you probably know more about this particular subject than your reader does. As a writer, you may have a different perspective on an issue than your reader has.

In each case, though, you must be careful to talk *to* your reader. You also have to be careful not to talk over your reader's head. To hit a happy medium between talking down to your reader and talking over your reader's head, you again need to reconsider your purpose in writing in the light of what you know about your audience.

Format

After you have considered purpose and audience, you are in a much better position to make choices about the most appropriate format. Perhaps you may decide that a phone call or a personal visit would be the best way to communicate. But if you decide that a written communication would be best because you want an accurate record, you then have further decisions to make regarding format. Should you write a letter, a memo, an informal report? Does the company have a preferred format for each? The choices you make will influence the reader's acceptance of your message. The format consists of the arrangement or layout of the message on the page. You want the format as well as the words to contribute to achieving your purpose with your audience.

If you are tempted to ignore the importance of format, pay closer attention to your junk mail. Those form letters we all receive are carefully assembled by experts to achieve the maximum impact on the reader.

The accompanying example, a four-page letter from the American Museum of Natural History, is typical. While the letter is admittedly long, notice how the writer has "pulled out all the stops" to guide the reader's eye as it scans the pages of the letter.

[3] Steven E. Pauley, *Technical Report Writing Today,* 2nd ed. (Boston: Houghton Mifflin, 1979), p. 6.

What's going on around here?

October 1985

Dear Friend:

What's going on around here?

Once dinosaurs roamed the earth ... migrating passenger pigeons filled the skies ... buffalo dominated the plains ... and early man crossed into North America over a Siberian land bridge.

Where are they today?

The American Museum of Natural History.

At the current rate of extinction, a fifth of all presently existing species — a million or more -- may have vanished from Earth by the end of this century.

Where is the only place you, and future generations, and -- critically -- future generations of scientists will be able to find them?

The American Museum of Natural History.

IF ...

1. We are able to continue adding new specimens to our collections.

2. We are able to preserve the incalculably rare, irreplaceable collection the Museum already has in its possession.

That is why I invite you to join the thousands of people who not only recognize the tremendous importance of the American Museum collection, but who also play an active role in the collection's expansion and preservation.

"Though exhibitions and events constitute the public face of the Museum, it has for the past century continued to be a research institution of international distinction."

(New York Magazine)

AMERICAN MUSEUM OF NATURAL HISTORY   CENTRAL PARK WEST AT 79TH STREET   NEW YORK, NEW YORK 10024

- 2 -

What places the American Museum among the very few top-ranking research institutions of its kind in the world?

OUR TREMENDOUS FACILITIES ... Labs, workrooms, storage areas, exhibition halls and offices in the 22 interconnected buildings covering 25 acres of land. The largest natural history museum in the world. The largest museum of <u>any</u> kind (<u>Guinness Book of World Records</u>).

SUPERB SCIENTIFIC RESOURCES ... 200 research scientists and research associates, among the most distinguished in the world. Hundreds of visiting scholars, scientists and students from around the globe. Loans and exchanges from our collection to some 3,000 scientific institutions and museums worldwide. The latest scientific equipment. The largest and most important natural science library in the Americas.

BUT ABOVE ALL, THE AMERICAN MUSEUM IS DISTINGUISHED BY ITS COLLECTION ... More than 39 million specimens and artifacts brought together as a scholarly resource over the course of the Museum's 116-year history — less than 1% of which are on public exhibition. Imagine if you can ...

. 16 million insects

. 8.5 million invertebrates

. 8 million anthropological artifacts

. 3.5 million photographic images

. 1 million bird specimens

. 1 million fish specimens

. 330,000 fossil invertebrates

. 250,000 mammals

. 230,000 amphibians and reptiles

. 385,000 library volumes

And if you think museums are supposed to be musty repositories only preserving the past, nothing could be further from the truth in our case!

<u>For it is these specimens that hold the answers to major questions that plague us today, new questions that mankind will continuingly face in the future.</u>

The Museum is involved each year in more than 300 research projects: studies that furnish scientists with answers to questions ranging from what role asbestos plays in pulmonary diseases to which species of kissing bug spreads a rare malarial disease ... from how fossils can help locate oil deposits on the ocean floor to how fish specimens can help trace the effects

- 3 -

of chemical pollution in a river ... from how we can stay the disappearance of
our rain forests and deserts and prairies, and the wildlife existing within
them, to baseline data that is the basis of scientific understanding.

If the American Museum is to continue fulfilling the crucial role it plays
in that kind of research, it is imperative to preserve its invaluable
collections and add to them! Once a species vanishes totally from our earth,
its secrets, potentially incalculable in value, vanish with it.

How does the Museum acquire new specimens?

1. From scientific expeditions. Today, still unknown and unrecorded
   biota exist in corners of our world -- still remote despite modern
   technology.

   A recent expedition to Neblina, a 6,000-foot mesa above the tropical
   rain forests of Venezuela, required eight days' travel by dugout canoe
   before scientists could even establish a base camp in this "lost
   world," where plants and animals have developed in almost total
   isolation. The expedition, still in its first year, has yielded a
   wealth of results, including a previously unknown "bearded" catfish
   caught in the Rio Mawarinuma. Of particular importance was that the
   genus of the species was also a new discovery.

2. Through gifts. At the present time a growing number of universities
   and other institutions are depositing their own scientific collections
   in the American Museum, recognizing the value in developing major,
   centralized systematics resources. Many private collections do as
   well.

   Unfortunately, many institutions are being forced to give up their
   collections for the very reason the American Museum is itself finding
   its resources so enormously strained: the increasing difficulty and
   cost of maintaining, managing, preserving and using these great
   "libraries" of material evidence.

We cannot simply "store" artifacts and specimens. The problems of erosion
and destruction so widely publicized in the world of art and antiquities are
infinitely multiplied within a museum of natural history. New ways and new
inventions must be utilized to preserve and protect a seemingly endless
variety of plant and animal substances, a multitude of man-made fibers and
artifacts.

The American Museum has been in the forefront of developing new methods of
conservation and preservation. But both the investigation and implementation
of such programs are costly beyond anything we could have imagined even a few
short decades ago.

As other institutions give up their collections, it becomes that much more
imperative that we be able to maintain our collection to the highest, safest,
most advanced standards possible.

- 4 -

That is why I issue this urgent appeal to you and others who can understand the importance of what we have done -- and what we must continue to do -- at the American Museum.

It is no secret that budgets are tight and funds short in many important areas these days. The funds we receive must go not only toward the expansion and care of our collections, but also toward the other valuable services our vast collection provides ...

  · exhibitions on a broad range of subjects in natural history and anthropology for more than 2.5 million visitors each year.

  · information generated by the Museum is disseminated to many millions of people worldwide through popular and scientific publications.

  · educational services such as demonstrations, lectures, film programs, performing arts programs, special courses and workshops for the community around us, aimed at people of all ages, with diverse interests and backgrounds.

The same New York Magazine article I quoted earlier noted that the American Museum of Natural History "speaks with a quiet voice ..."

THE TIME HAS COME WHEN WE MUST RAISE OUR DECIBEL LEVEL A LITTLE!

If the Museum is to continue serving as a unique resource for future generations, through which they can see and study and come to understand the life on earth, much long vanished, we must maintain our collection, and we must find the funds to do so.

We turn to you to help us do so.

Please give us your support by making a tax-deductible contribution of $25, $50, $100, $500 — whatever you can give us to support this vast and significant undertaking.

If you share our belief that our best hope for the future lies in our study of the present and the knowledge we have gained from the past, please help us carry on with this critical task.

Sincerely,

T. D. Nicholson

Thomas D. Nicholson
Director

P.S. In gratitude for a contribution of $50 or more, I will send you a copy of the 116th Annual Report of the American Museum of Natural History. This handsome, illustrated volume will give you a broad view of the scope and diversity represented in our scientific work and of the public services we make available also. I know you will enjoy and admire it.

Reproduced with the kind permission of the American Museum of Natural History.

Like the writer for the museum, you can use a variety of signals in your writing to help your reader's eye move down the page. Such signals include:

Vertical spacing.
Horizontal spacing.
Indentation.
Underscoring.
Capitalization.
Enumeration using numbers, bullets, or letters.
Headings.

Evidence

Although evidence sounds as though it belongs in a courtroom, it is the meat of any piece of writing just as it is the meat of a lawyer's case. If you try to write before you have sufficient information, you will find the process difficult and the final product empty and unconvincing.

What is evidence? For a writer, evidence begins with facts and information gathered through observation, interviewing, or research. If you are trying to inform or persuade your readers, you must give them facts or specific information that supports your purpose and clarifies your meaning. *Telling* the reader is one thing; *showing* the reader the evidence supporting your position is another.

Organization

Readers detest the feeling of confusion they experience when they can't find the point and can't see any plan in what they're reading. You as the writer may think you are being very organized in your presentation, but the reader may not be able to follow your plan because it is not designed for readers. Your plan is part of your effort to make sense of the subject for yourself rather than for your reader. In the useful terms suggested by Linda Flower, the problem is that the organization is *writer-based* rather than *reader-based*.

Suppose, for example, that you must file a progress report each week. The purpose of the report is to let your boss know that the project you are working on is coming along—whether or not you are running into any problems she ought to be aware of. You find the report a time-consuming pain in the neck, so you sit down at the typewriter or word processor and bang out a hasty account that tells what you did last week and when you did it and who said what to whom. You present a narrative that begins with Monday and ends with Friday. Are you filling the needs of your audience? Wouldn't another organizational pattern other than chronological be more useful, given the purpose of the report and the needs of the audience?

To give you a better idea of how a PAFEO approach to the problem of writing can make a difference, let's look at an example of a letter written without the least consideration for the concerns represented by the PAFEO acronym.

United States of America
Office of
Personnel Management

Pittsburgh Area Office
Federal Building
1000 Liberty Avenue
Pittsburgh, Pennsylvania 15222

Dear Mid-Level Eligible:

This is to inform you that as of June 30, 1980, the competitor inventory established under the terms of our Mid-Level Recruiting Bulletin, PP-5-02 will be terminated. This means that your name can no longer be referred for consideration in filling Mid-Level positions with Federal agencies based on your current eligibility after that date.

The reverse side of this letter explains the new procedures through which agencies will fill competitive vacancies for Mid-Level positions. Therefore, if you continue to be interested in applying for such positions, you must follow the new procedure in so doing.

This letter was sent to your mailing address in effect at the time you initially established eligibility. There was no capability of subsequent updating your address, in the event of changes, on our automated record; however, any address changes reported by you were manually recorded on your application file in this office.

Your interest in Federal employment is appreciated.

                    Sincerely,

NEW PROCEDURES FOR FILLING ''MID-LEVEL'' POSITIONS
IN THE MID-ATLANTIC REGION

As of July 1, 1980, new procedures are being used to fill Mid-Level positions (GS-9 through GS-12 positions not filled through other examinations) in Pennsylvania; Maryland; Delaware; Virginia; West Virginia; Camden County, New Jersey; and Belmont, Lawrence, and Jefferson Counties, Ohio. Mid-Level Recruiting Bulletin PP-5-02 is no longer used and the Pittsburgh Area Office of OPM is no longer keeping a list of eligibles for Mid-Level positions.

Instead, Federal agencies have been delegated authority to publicize their own ''Mid-Level'' vacancies, accept and evaluate applications, make selections, and return the applications of those who are not qualified or not selected. No on-going lists of eligibles will be maintained by the agencies. Applicants will file separate applications for each vacancy announcement in which they are interested.

Agencies are required to post their ''Mid-Level'' vacancy announcements in the Federal Job Information Center and State Job Service offices in the state where vacancies are located. They are also required to make selections in accordance with the same laws and regulations followed by OPM.

OPM will still examine for Mid-Level vacancies for the few agencies that have not accepted the delegated authority. The OPM area office in the state where such a vacancy is located will publicize the vacancy in its Federal Job Information Center and the State Job Service offices, accept and evaluate applications, refer the best qualified applicants to the agency, assure that all selections are proper, and return the applications of those who are not qualified or not selected.

These new procedures were devised primarily to assist agencies in filling their ''Mid-Level'' vacancies more quickly. There are also advantages to applicants. Only actual vacancies are publicized, rather than general occupational categories, so applicants know exactly what positions they have applied for and exactly what happens with each of their applications. Also, applicants may now apply to, and be considered by, several agencies simultaneously.

Those interested in ''Mid-Level'' opportunities should maintain periodic contact with the Federal Job Information Center or a State Job Service office in the state where they wish to work in order to stay apprised of current vacancies.

The sample is a form letter sent out by the United States Office of Personnel Management. There is nothing wrong with sending a form letter, but writers of form letters have to keep in mind the needs of a much wider audience. In this case, the writer knew that the letter had to be directed to people not yet working for the government, people who would probably be unfamiliar with some or all of the processes involved in securing employment with a government agency. (In its orginal form, the letter appeared on both sides of the same sheet of paper.) Reread the letter carefully. Then using the form we provide on p. 21, write a PAFEO analysis of the two-page letter.

FORM FOR APPLYING THE PAFEO PROCESS TO A WRITTEN MESSAGE

Purpose (Being as specific as possible, what is the goal the message is working toward? Does the message have more than one purpose? What is the reader supposed to do or think after reading the message?)

_____

_____

_____

_____

Audience (Who is the reader? Is there a primary and a secondary audience? What does the reader need to know? Is the message too specialized for the reader? Does the message belabor the obvious for the reader?)

_____

_____

_____

_____

Format (Does this particular message lend itself to a specific format? Are sentences and paragraphs easy to follow, or do they confuse the reader because they are too long and too complicated? If the messages fall into different parts, should these parts be formatted differently?)

_____

_____

_____

_____

Evidence (Does the message tell the reader everything the reader needs to know in a way the reader can understand?)

_____

_____

_____

_____

Organization (How should the message be organized to make its evidence as accessible as possible to the reader?)

_____

_____

_____

_____

After you have written your PAFEO analysis, take a look at the following sample analysis written by one of our students. You may also want to compare yours to those written by your classmates.

You will see, we trust, how the writer could have made this letter more effective by using the PAFEO process. We hope that this experiment will demonstrate how much PAFEO can help you in your efforts to make your own writing decisions by focusing on the problems you must solve for your reader.

ONE STUDENT'S PAFEO ANALYSIS

Purpose: The message gives some mixed signals about purpose. The first page, addressed to "Mid-Level Eligible," seems to want to tell such a person about changes in procedures by which Federal jobs will be filled. The writer would want then to persuade the reader to reapply under the new system by stressing the advantages the new system holds for the reader.

The second page of the message seems to have a different purpose. The heading details new procedures for *filling*—not *applying for*—mid-level positions.

Audience: The audience at first seems to be people not yet working for the Federal government who need to know to reapply for Federal jobs. Such an audience will have problems with *mid-level eligible, competitor inventory, Mid-Level Recruiting Bulletin,* and even *OPM.*

The audience for the second page of the message seems different. This audience includes people who will be filling the announced vacancies, people already working for the Federal government who are charged with filling competitive, and presumably, noncompetitive vacancies.

These two audiences have different needs and come to the letter with different levels of preparation.

Format: Either audience will be helped by short sentences and short paragraphs.

The letter's second paragraph promises a set of procedures, but it is difficult to find a list of steps on the second page. There are six paragraphs, but each does not translate into a step in a procedure.

Evidence: The reader needs to know that there is a change effective July 1. The reader needs to know that the change means everyone still interested must reapply. The reader needs to know the advantage of the new system to himself or herself. The reader needs to know where to go for further information. The reader needs to know what agencies follow the new procedures, what agencies don't. The reader needs to know what happens to applications once they are filed.

Organization: An approach that moves from general to specific or less complex to more complex might work here. The letter could, for instance, begin with the announcement of the change, move on to explain the benefits of the new system, detail the steps of the new application procedure, and follow with any other necessary background information.

Exercises

1. When customers fail to make payments on time or when, for a variety of reasons, payments are not received or recorded, businesses adopt different strategies to ensure that payments due are collected. Some businesses ignore the situation and bill customers again in a month. Others send interim reminders. In extreme cases of default and nonpayment, businesses turn, of course, to collection agencies or to the courts.

   Letters A through C attempt to jar the memories of customers who owe payments now overdue. Letters A and B are first notices; letter C is a second notice. As an exercise for classroom discussion, use the PAFEO process to analyze these letters and, where appropriate, to suggest ways of rewriting these letters to improve their effectiveness.

LETTER A

CHARLES OF ST. LOUIS
Fine Men's Apparel
One Hundred Broadway
St. Louis, Missouri 63102
314-444-9999

October 16, 1986

David Landry
P.O. Box 97
Osage Beach, Missouri 65065

Account #: 34-DL65065-1

Dear Customer:

When your account was opened, you agreed to make regular monthly payments based upon your contract terms.

Payment is now overdue in the amount shown below. Please send your remittance promptly to bring your account up to date.

Charles of St. Louis
Department of Accounts

Amount Due: $40.00

LETTER B

MAUI FEDERAL TRUST
Nohea Kai Drive
P.O. Box 97
Lahaina, Hawaii 96764

May 27, 1986
RE: # 010-57686765432-9087632

Dear Customer:
  We have not received payments in accordance with your
Installment Sale Contract.

  We urge you to use your bank credit wisely and submit
your delinquent installment today.

Sincerely,

Consumer Lending Department
808-777-8999

LETTER C

Dear Friend of GQ:

Perhaps it's just a case of "putting it off" as so many of us
are apt to do -- that is, sending in the amount due for your
GQ order.

Or, perhaps you didn't receive my first reminder letter to
you. In either case, won't you let me hear from you?

A convenient reply envelope is enclosed for your remittance...and thanks for cooperating.

Cordially,

M. A. Pfenninger
for GQ

18-1109

Reproduced by permission of the publisher, Condé Nast Publications, Inc.

2. A failure to deliver services or goods promised requires an especially tactful and effective written follow-up. Here is an example of such a follow-up from a public transportation system in a large city. As an exercise for classroom discussion, use the PAFEO process to analyze this message and to suggest ways of rewriting the message to improve its effectiveness.

AN EXPLANATION

On Tuesday, April 21, at 724 AM at First and Main Streets, a device failed that provides electric current to control subway train movements. The result was an inability to display signals at Second and West Streets and at First and Main Streets and to operate track switches at First and Main Streets.

Consequently, all downtown subway train movements came to a halt. By 738 AM supervisory personnel had been stationed at strategic locations. So, subway train movements were resumed. But subway trains had to be operated on a one-at-a-time basis between Second and West and First and Main Streets. This decreased normal train frequency, to one about every six minutes instead of one about every three. At 851 AM the defective equipment was repaired, and operations returned to normal.

As you discovered, the above events messed-up the busiest part of the morning rush hour, completely! We know you suffered aggravating delays. We regret your inconvenience. However, our operations during the equipment failure were designed to assure your safety. Thank you for your patience and understanding.

The City Subway System
4/22/85

3. Your school's governing board has just authorized a 10 percent increase in tuition and fees for the coming academic year. Your school's president must notify students and their parents about this increase. Use the PAFEO process as the basis for classroom discussion of how the president should prepare to write this letter.

Here are considerations to use in your discussion. Since the president needs to notify parents and students, should the president send separate letters to students and parents? Tuition increases won't do much good if these increases result in fewer students attending your school. How can the president assure readers that education is worth the added cost?

Any number of factors could be contributing to this increase. Use any of the following that you find helpful, imagining for the purposes of this exercise that they are true for your school.

- Rate of return on the endowment has been steady but has not increased as much as projected.
- Utility costs campuswide have skyrocketed.
- To retain faculty, the school must allow faculty some kind of annual increase in salary. In past years, that increase has been 5 percent. To retain and attract good faculty, an increase of at least 5 percent must be offered again next year.
- In the long run, your school will save substantially by computerizing as many offices and functions on campus as possible. In the short run, such computerization requires a substantial outlay of funds.

- Federal and state cuts in aid to education have caused your school to assume an increasing share of the burden of providing students with financial aid.
- In the past few years, your school has tried as much as possible to defer maintenance around campus, but even with a careful maintenance plan, repairs need to be made, walls need to be painted, and facilities need to be upgraded.
- Faculty aren't the only people on campus who deserve salary increases. Your school must be fair to administrative and support personnel at all levels. These people also expect at least a 5 percent salary increase next year.
- A selective hiring freeze has been in effect, but it must soon be lifted so that problems affecting the quality of education and service your school provides don't arise.

If your instructor directs, you might try writing a version of the president's letter based on what you have learned in your classroom discussion.

4. Three examples of actual business messages follow—a sales letter, a memo, and an information notice.

Use the PAFEO process to analyze these messages and to help you rewrite them to improve their effectiveness.

SAMPLE A

THE GREATER FOND DU LAC DAILY JOURNAL
100 North County Line Road
Fond du Lac, Wisconsin 54934

December 5, 1987

Terry Pryor
Sales & Advertising Manager
Pryor Lumber and Hardware
3500 Lake Road
Fond du Lac, Wisconsin 54935

Dear Terry:

I know that this is an extremely busy time for you, so I'll keep this note brief.

I've been trying for some time to arrange a mutually convenient appointment to meet with you. Based on various market changes, including new Pryor store locations, our newspaper's readership and demographic changes, and increased use of our advertising space by your competition, I feel it would be extremely beneficial for you and us if we could meet and discuss some of these matters by the end of the year, or early next year.

Please let me know when you'll be available, and I'll arrange my schedule to fit yours. You can reach me at 555-8990.

Thanks for your time and consideration.

                                    Sincerely,

                                    Marty

                                    Marty Finkelstein
                                    Advertising Sales Manager

MF/bc

P.S. Enclosed is a copy of today's edition for your inspection.

SAMPLE B

                                    November 2, 1987

TO: The Campus Community

Dear Colleague:

  The very successful introduction of the Deferred Payment Plan for tuition and the resulting increase in payment activity in the Bursar's Office, combined with the physical limitations of the cashier area (only 2 windows) is causing an already bad situation to become worse. Until, at some future time when we might find more space, the only way to provide some relief is to search for ways to reduce the need to transact business in this office. With this objective then, effective November 19, 1987 the sale of postage is being transferred to the Duplication and Mail Department. After that date the Bursar's Office will no longer sell stamps.

Thank you for your cooperation and understanding.

                                    Sincerely,

                                    Maria Hernandez-Cortez

                                    Maria Hernandez-Cortez
                                    Vice President, Business Affairs

MH-C/dcf

SAMPLE C

Information Notice   Department of the Treasury   **Internal Revenue Service**   │   Number: 83–18

Address File for Forms W-2, Wage and Tax Statement

Beginning in pay period 1, December 25, 1983 to January 7, 1984, the Payroll/Personnel System employee address file maintenance processes will be revised.

A listing containing employee address information as of pay period 23, October 16-29, 1983, will be prepared by the Data Center and forwarded to your office for review, correction, and return to the Data Center. The Data Center will update the address information with the required corrections prior to distribution of the Forms W-2, Wage and Tax Statement, for 1983.

Beginning in pay period 1, all employee address file additions and changes will be accomplished via submission of the new Form 8147, Employee Address and Mail Authorization. Form 8147 will be completed by all new employees when they enter on duty with the Internal Revenue Service, and all other employees whenever their address information changes. Form 8147 can be obtained from your Personnel Office; completed forms must be submitted to your Personnel Office for processing.

The information that you will be providing on this form will be used to establish and maintain address information needed for the preparation on your Form W-2. In the future this form will also be used to direct the mailing of your Leave and Earnings Statement and salary check. You will be notified in advance when each of these future options becomes effective.

Note: The above instructions amend and supplement IRM 0290.3, Payroll/Personnel Processing Handbook, and will be reissued in accordance with 420 of IRM 1230, Internal Management Document Systems Handbook, by a manual transmittal in clearance.

5. Your state legislature has just passed and sent on to the governor for signing into law a bill that imposes a state-wide school tuition tax. This tax represents a $250 a year surcharge to be paid by all students attending postsecondary schools in your state. The tax applies to both resident and nonresident students. Using the PAFEO process, write a letter to the governor asking him to veto this bill.

6. Return to the letter from the Office of Personnel Management (pp. 19–20). Review our discussion and your own PAFEO analysis of that letter. Then using the PAFEO process, rewrite the letter.

In rewriting the letter, assume that:

- Its primary purpose is to alert applicants to the new procedures.
- Its secondary purpose is to get applicants to reapply.
- Its third purpose is to provide any necessary additional background information.

Also assume that the new procedures apply to all federal agencies except the State Department, the Internal Revenue Service, and the National Parks Service.

The Office of Personnel Management will continue to screen applicants for positions with these three agencies.

As an attachment to your letter, assume there is a pamphlet detailing "the laws and regulations followed by OPM" in filling job vacancies with federal agencies.

CHAPTER 2

Purpose and Audience

In Chapter 1, we introduced you to the PAFEO process by explaining the close relationship between your purpose in writing and your audience. In this chapter, we intend to show you how you can systematically define your purpose and analyze your audience.

You may be tempted to take the purpose for granted, especially if it seems obvious. If you do so, you are likely to be satisfied with a statement of purpose that is too general to help you write more effectively. You may also end up with a statement of purpose that will mean less than you think it will to your audience.

Composition textbooks divide writing according to whether its purpose is narrative, expository, or descriptive. Using such broad categories as these, we find some overlapping. A narrative piece can contain much that is descriptive. Expository prose can introduce a short section of narrative to make a point clear. Purpose does not always exist in a pure state.

Here's an example that illustrates the complex, multipurpose nature of many business communications. As a member of a consulting firm, you make a proposal to a prospective client. The proposal must meet two important purposes:

1. It must sell your firm's product or services.
2. It must spell out the terms of what your firm is contracting to do.

Although these two purposes may be tops in priority, the reality of your proposal can involve other purposes that must be considered in the writing choices that you make. These other purposes might include:

- The desire to impress your boss with your handling of this proposal so that you will be considered for promotion.
- The need to avoid troubles and delays from the legal department.
- The necessity for anticipating future problems that your client might have with public relations for your proposal.

We might call all of these *operational purposes,* and it is just as important to identify such purposes as it is to identify those to which we give top priority. The effectiveness of your writing choices and the eventual success of your proposal hinge to a great extent on how well you can clarify your many purposes.

A Systematic Approach to Determining Purpose

There are only four basic purposes for writing:

1. To *clarify* what the subject is (to define a word, explain a concept, report evidence).
2. To *substantiate* a thesis about the subject (to demonstrate that a conclusion or an inference about the subject is correct).
3. To *evaluate* the subject (to judge the subject as good or bad with reference to useful, ethical, or aesthetic standards).
4. To *recommend* that something be done about the subject (to persuade people to think or act differently about the subject).[1]

These general aims pertain to what you intend to do with the *subject.* But, as we have seen from the example above, when you write a complex business communication, you may have many other purposes that affect the choices you make in writing. You may have to achieve conflicting purposes and respond to the needs of different audiences with different expectations—all at the same time.

Suppose, for example, that you must write a bad-news letter denying credit to a long-time customer. Your basic purpose may be to evaluate the customer's worthiness for credit and to recommend acceptance or rejection. But the actual letter you write must serve all of these operational purposes as well:

1. Inform the customer of the bad news.
2. Explain the reasons for the denial of credit.

[1] We borrow this helpful four-part classification from Caroline Eckhardt and David H. Stewart, "Towards a Functional Taxonomy of Composition," *College Composition and Communication,* 30 (December, 1979), 338–342.

3. Persuade the customer of your company's continuing good will.
4. Cover any legal requirements.
5. Satisfy your supervisor.
6. Provide a record of the facts.

In our experience, we have found that students are tempted to oversimplify their purpose statement, dealing only with the general aim and giving insufficient thought to the operational purposes. When asked what their purpose is, they quickly reply, "To inform," or "To recommend." But more is needed if the writer is to make the most effective choices regarding organization, evidence, and format.

What is needed is a systematic approach to purpose that will help writers see what problems they must solve for each of their readers, thus enabling them to write a purpose statement that will help them as well as their readers. Such an approach has been developed for technical writing by J. C. Mathes and Dwight Stevenson.[2] Although their method is most useful in report writing, you may find it useful in helping you with certain kinds of memos and correspondence, especially those of a trouble-shooting nature.

Since Mathes and Stevenson directed their approach to the needs of engineers, we have adapted their ideas to make them fit the problems of general business writing. In doing so, we have chosen to present the steps as a series of questions:

1. What is the problem in the organization that created the need for this message? To put it another way, why are you being asked to write this message?
2. As a problem-solver, which of the four general aims of writing apply to the problem at hand? (Remember, there may be more than one.)

   - Clarifying the problem by defining, explaining, or reporting evidence.
   - Substantiating a conclusion about the problem.
   - Evaluating the problem, judging it by some appropriate standard (e.g., advantageous or disadvantageous to the company, feasible or impractical, cost effective or not, ethical or unethical, legal or illegal).
   - Recommending appropriate action.

3. As a writer, what do you want your readers to think or do as a result of reading your message?

If you can answer these questions about purpose, you will know *what* you have to do and *why*. The questions will help you see

   - the problem in its original context,
   - your purpose as a problem-solver, and
   - your purpose as a writer who must communicate to diverse readers what they need to know about the problem.

Answering these questions will provide you with a systematic approach to writing a good purpose statement. Such a purpose statement, placed at

---

[2] J. C. Mathes and Dwight W. Stevenson, *Designing Technical Reports: Writing for Audiences in Organizations* (Indianapolis: Bobbs-Merrill, 1976), pp. 24–42.

or near the beginning of your message, will improve your own and your audience's understanding of the information that follows.

Every reader has values, interests, prejudices, and particular viewpoints. Let us now turn to some suggestions on how you can approach the complex problem of meeting the needs of varying audiences.

A Systematic Approach to Analyzing Your Audience

Let's begin our discussion of audience analysis by listing three questions you will want to ask about your audience each time you sit down to write. These questions have value whether you are writing a letter, a memo, or a report:

1. Who's your audience?
2. What do you know about your audience?
3. What do you want your audience to do or think after reading your message?

Now let's expand upon the implications of each of these questions, keeping in mind how they relate to our earlier discussion on determining purpose.

Who's Your Audience?

If your audience is all the registered voters in the 86th Congressional District, you will need to compose your message differently than if your audience is A. K. Rosenberg, President of the First National Bank. Explaining the prime rate to someone with an MBA is one thing. Explaining the prime rate to a layperson is another. The prime rate is still the prime rate, but the audiences differ greatly in their abilities to understand technical information.

In answering this first question, be specific. Try as much as possible to get an accurate mental picture of your reader or readers before you write. Such a mental picture is admittedly easier to come up with when writing to A. K. Rosenberg than it is when writing to all the voters in a congressional district. You can readily imagine A. K. Rosenberg sitting across the desk from you. You can, however, still force yourself to concentrate on as large and as diverse an audience as all the voters in the 86th Congressional District by continuing your audience analysis with the second of the three questions listed above.

What Do You Know About Your Audience?

First, your audience is composed entirely of voters. Second, they may be overwhelmingly Democrats, or Catholics, or blue-collar workers, or first-generation Americans. Third, they may share common concerns. They may all worry about unemployment, toxic dump sites, a shortage of rental housing, and skyrocketing automobile insurance rates.

At the same time, you may know something specific about A. K. Rosenberg which you can use in writing to her. She may be the president of a large commercial bank or a small savings and loan. She may be a veteran bank president or a recently promoted chief executive. She may have a background in international banking, or she may have previously headed up a commercial lending department.

In asking your second question about what you know about your audience, you are really asking what you know about your audience's background. When writing to an individual, you need to assess as much background information as you can gather. Consider just the following three questions designed to elicit background information:

1. Are you writing to a new or to a long-standing customer?
2. Are you writing to someone with a Ph.D. or to someone who dropped out of grade school?
3. Are you writing to a boss who wants everything spelled out in tedious detail or to a boss who wants everything summarized on one typed page?

Under most circumstances, the latter has what seems the more sensible and efficient way of dealing with information, but you cannot afford to offend the boss who wants all that tedious detail without putting your job in jeopardy. Again, keep asking yourself specifically what you know about your audience and that audience's background. Then use what you know to check yourself as you write and rewrite your message.

So far we have concentrated on *personal data* that could have bearing on your audience's receptivity to your purpose. Here are some issues related to *professional data* that you may also need to consider in a systematic analysis of your audience.

What's your audience's
   education and training?
   knowledge of your purpose in writing?
   knowledge of your interests?
   knowledge of your responsibilities?
   interest in or involvement with your purpose in writing?
   understanding of your purpose in writing?
Does your audience
   hate details?
   demand details?
   feel intimidated or threatened by you?
   feel superior to you?

Because you must consider the impact that your message may have on your audience, you must answer a third question as you continue to analyze your audience systematically.

What Do You Want Your Audience to Do or Think After Reading Your Message?

This third question reintroduces you to your purpose in writing. Presumably, you do have a specific purpose in writing to your audience. Your audience may, however, read with a different purpose in mind than you intended.

For instance, you may want to sell widgets, but your audience may want to know all about widgets before reading a sales pitch. Indeed, you may find that some initial explanation clearly directed to the audience's interests, needs, and background might go a long way toward helping you sell a great many widgets.

You need to resist the temptation to get high-handed in your message. You may be really annoyed about something. You may be in a position to play boss and put your foot down. But in doing so, you run the risk of making a fool out of yourself and of failing to accomplish what you set out to do. Keep your purpose clearly in mind, but keep your audience and that audience's possible response or reaction to your message just as clearly in mind.

Consider the following situation and the actual written response that came out of that situation. A new manager, aged 25, is hired to supervise a billing department staffed by fifty billing clerks. Each clerk has worked for the company for at least ten years. All the clerks are older than the new manager; some are old enough to be one of the manager's parents.

The new manager, after only two weeks on the job, sends the following memo around to all the billing clerks:

September 14, 1986

To: All Billing Clerks
From: Louise Cecculli, Manager

EFFECTIVE IMMEDIATELY: NO MORE EATING AT DESKS!

Cecculli's memo is going to cause problems; she runs the risk of sounding unprofessional. There may be a variety of reasons why the billing clerks should not eat at their desks. The bills may have been coming through stained with mustard, relish, or coffee. The computer may be rejecting punch cards because they have been sticking together thanks to peanut butter-and-jelly smudges.

Cecculli wants to end these problems. They are genuine problems affecting departmental productivity and efficiency. Cecculli's memo, however, may antagonize her audience, fail to accomplish her purpose, or create new problems.

The clerks are, first of all, adults. They may snack at their desks because it is convenient to do so. They may even eat their lunches at their desks to catch up on or get ahead with their work. Cecculli's memo takes none of these possibilities into consideration. Instead, Cecculli decides to pull rank. She's the boss; her word is law.

Here's a suggested revision of Cecculli's original memo. This second memo is designed to solve the problems of sticky computer cards and coffee-stained bills, without creating new problems. There is no guarantee that the second version of this memo will solve all the problems that eating at desks causes. Additional reminders in person may be necessary, but the first version of the memo will do more harm than good. It will generate hard feelings. The clerks may well view the new manager as an upstart, and the problems of sticky punch cards and stained bills will doubtlessly continue.

```
September 14, 1986

To:       All Billing Clerks
From:     Louise Cecculli, Manager
Subject:  Problems Caused by Your Eating at
          Your Desks

I need your help in solving a problem that is affecting
our department's overall productivity and efficiency.

Lately, the computer has been jammed by coded billing
cards that have become stuck together by what appears to
be grape jelly. In addition, copies of bills for file have
come through illegibly because of coffee and other stains.

Please don't eat or drink at your desks, and please make
sure that you wash your hands after returning from coffee
breaks in the lounge or from lunch in the cafeteria.

We all already have enough headaches that we can't do
anything about, but here's a case where we will all be
helping ourselves out.

I appreciate your cooperation.
```

Since your audience will, more times than not, be made up of reasonable people, attempt to look at issues or problems by considering the effect your message will have on your intended audience.

- Will your message motivate your audience to respond the way you want your audience to respond?
- Will your message antagonize your audience?
- Will your message bewilder your audience?
- Will your message insult your audience?

Such questions can be answered specifically by considering the kinds of audiences in business and industry you routinely address, the special problems these audiences pose for you, and the kinds of needs these audiences have.

The discussion that follows categorizes typical audiences for letters, memos, and reports from the world of work.

The Kinds of Audiences You Can Expect

In some ways, every writing situation is unique. It is, however, possible to delineate some general categories of readers and to anticipate the different needs the readers within each category may have:

- The general public or the layperson.
- Subordinates or clients.
- Bosses or superiors.
- Multiple or mixed audiences.

There will doubtlessly be some overlapping, but these categories do give a fair overview of who is out there as a potential audience for your letter, memo, or report.[3]

The General Public or the Layperson

The memo to all employees, the letter to "Occupant," the personalized form request for a donation to a worthy cause, or the general announcement of a new service or product are all addressed to the general public or to the layperson.

Writing to such an audience requires careful planning. Generally, the two most important words you as a writer can use to communicate with your audience are *you* and *your,* since you need to direct your message *at* your audience.

> The EXODIET program combines a routine of diet and exercise for you carefully designed to meet your body's needs.

A good way to begin messages to readers in this category would be to stress the familiar, use an analogy, or define the unfamiliar.

> Our new Individual Retirement Accounts (IRAs) combine a number of features you are already familiar with if you have a savings account or a company-sponsored pension plan.

The general public or the layperson needs ideas that are expressed in plain, simple language familiar to them. They appreciate shorter sentences

---

[3] The pioneering work on audience analysis for business and technical writing was done by Thomas Pearsall in *Audience Analysis for Technical Writing* (Beverly Hills, CA: Glencoe, 1969). Our four categories for audiences represent a conflation of materials Pearsall originally developed.

(15 to 25 words) and shorter paragraphs (6 or fewer sentences). Further, they appreciate examples, asides, and comments directed specifically at them.

Your approach throughout a message aimed at the general public or the layperson should be to tell your readers a story they will want to read. The following letter urging readers to increase their insurance coverage is a good example of a message directed at the general public or the layperson.

NOW . . . FOR A LIMITED TIME
YOU CAN INCREASE YOUR COVERAGE
WITH A GUARANTEE OF ACCEPTANCE!

Dear Policyholder:

Have you given serious thought to how continuing inflation affects the value of your life insurance? Nearly every month the cost of living goes up while the buying power of every dollar goes down. The price of food, clothing and day-to-day household operations--the largest expenses for most families--has soared.

If your family tragically lost you today, your present life insurance coverage might not be capable of doing the job you intended it to do . . . because inflation has taken away the buying power your insurance once had.

That's why it's time for you to consider updating your present coverage under the Amalgamated Intercounty Workers Group Term Life Insurance Plan to provide the security your family needs. Today, you have the opportunity to increase your protection easily . . . and affordably!

TO INCREASE YOUR COVERAGE by an additional $12,000, complete the Application attached to your premium notice and mail the entire form along with a check for the semi-annual premium (see amount indicated on the premium notice). That's all you need to do!

IMPORTANT . . . if you apply by June 1, 1986, you are guaranteed this additional coverage if you can answer ''NO'' to three short questions.

Take advantage of this easy, convenient way to increase your life insurance coverage . . . and keep pace with today's economy!

Sincerely,

Subordinates and Clients

In writing to subordinates and clients, you need to watch your tone, the attitude you take toward your reader. Diplomacy is the key here. A careless message can lose a customer or cause problems in labor relations. You'll need to make sure that you are clear, concise, and tactful. Louise Cecculli's message earlier in this chapter was clear and concise—"Effective Immediately: No More Eating at Desks"—but it lacked tact.

In addressing subordinates and clients, you also need to balance your wants with their needs and capabilities. There is nothing wrong with saying "I," "me," "we," or "us," but remember your reader too. That reader responds better to the word "you." When your message benefits the reader, stress that benefit first.

> These new procedures will help you the customer keep better and easier track of your payments. They will also reduce the amount of paperwork we have to deal with.

NOT

> These new procedures will reduce the amount of paperwork we have to deal with. They will also help you the customer keep better and easier track of your payments.

Bosses and Superiors

You must be especially careful not to waste your reader's time when you are writing to your boss or superiors. Again, tact will be important, but you will also need to be politic.

You will need to supply your reader with adequate detail. You will need to separate opinion from fact and judgment. The bottom line is important to readers in this category. There are, however, two organizational patterns you can follow for arriving at the bottom line depending on what your reader needs from your letter, memo, or report.

Rändi Sigmund Smith calls these two organizational patterns the *logical* and the *psychological*. If the facts are more important to your boss or superiors than your analysis of those facts, you can use a logical organizational pattern:

· Introduction and statement of the problem.
· Data.
· Discussion or analysis of data.
· Conclusions.
· Summary with recommendations.

If, however, your analysis of the facts is more important to your boss or superiors than the facts themselves, you can use a psychological organizational pattern:

· Introduction and statement of the problem.
· Conclusions and recommendations.

> • Discussion or analysis of data.
> • Summary.
> • Appendix with data.[4]

Your boss or your superiors need information in a useable form. If visuals will help, supply them. If an opening executive summary or abstract will be of value, write one. Assess your reader's needs and meet them.

More than twenty years ago, Westinghouse Corporation conducted a study of the reading habits of its managers. The results of that study still have value today. They indicate a pattern in the way managers read reports that has a bearing on the way you need to write and organize messages— be they letters, memos, or reports—directed at your boss or superiors.

> The kind of information a manager wants in a report is determined by his management responsibilities, but how he wants this information presented is determined largely by his reading habits. This study indicates that management report reading habits are surprisingly similar. Every manager interviewed said he read the *summary* or abstract; a bare majority said they read the *introduction* and *background* sections as well as the *conclusions* and *recommendations;* only a few managers read the *body* of the report or the *appendix* section.
>
> The managers who read the *background* section, or the conclusions and recommendations, said they did so ". . . to gain a better perspective of the material being reported and to find an answer to the all-important question: What do we do next?" Those who read the *body* of the report gave one of the following reasons:
>
> **1.** Especially interested in the subject;
> **2.** Deeply involved in the project;
> **3.** Urgency of problem requires it;
> **4.** Skeptical of conclusions drawn.
>
> And those few managers who read *appendix* material did so to evaluate further the work being reported. To the report writer, this can mean but one thing: If a report is to convey useful information efficiently, the structure must fit the manager's reading habits.[5]

Multiple or Mixed Audiences

There will be times when your audience will consist of a mix of the categories we have just discussed: laypeople, subordinates and clients, the boss, and the boss's boss.

Format and organization can be the keys to dealing with mixed or multiple audiences. A nineteenth-century American engineer hit upon a sensible solution to the problem of audiences with different levels of interest or expertise. Arthur M. Wellington had different parts of his *The Economic Theory in*

---

[4] Rändi Sigmund Smith, *Written Communication for Data Processing* (New York: Van Nostrand Reinhold, 1976), pp. 178–179.
[5] Richard W. Dodge, "What to Report," *Westinghouse Engineer,* 22 (July–September, 1962), 108–111. The full text of this article is reprinted as part of the exercises at the end of Chapter 10.

the Location of Railways set in three different typefaces. He used large type for material designed for the lay reader. He used medium-sized type for the reader with some technical facility. He used small type for the reader who wanted detailed scientific information.[6]

Wellington's typographical solution to the problems of multiple audiences is copied with slight variation by writers of some of today's corporate annual reports. The letter from the president specifically addressed to the most general audience is printed on light color paper. The body of the report, which is of special interest to the stockholders, is printed on medium color paper, while the financial statements and analyses, most comprehensible to auditors and CPAs, are printed on a dark paper.

While you may not want to vary type face or page color, you can, nonetheless, key different sections of your message to different audiences by judiciously using captions and headings to separate clearly the parts of your message:

Abstract or executive summary.
Introduction.
Body or data.
Recommendations.
Conclusions.

You can also progress from general to specific or from simple to complex. In doing so, you will assure that you decrease the size of your audience only as you increase the difficulty of the material you present. If you were to begin with the most complex, you would more than likely lose most of your audience immediately. The headings or captions will, in turn, allow the busy expert to jump ahead to whatever he or she needs to read.

Tone

In passing, we have already mentioned tone, the attitude you take toward your subject and your reader. You can vary tone depending on the writing situation. The more formal the document, the more formal you will want your tone to be.

In dealing with any audience, you will want to adopt a tone that will help you achieve your purpose. Such tones can be forceful, polite, personable, helpful, and confident.

In establishing tone, you should avoid accusations and temper tantrums. You should also resist the temptation to play the boss or know-it-all. Phrases such as "as you probably know" or "needless to say" can set the wrong

[6] Arthur Wellington, *The Economic Theory in the Location of the Railways,* 2nd. enlarged edition (New York: Wiley, 1887), p. viii. The history of technical writing suggests that the best writers have always tried to accommodate their prose to their reader's needs and abilities. See Walter James Miller, "What Can the Technical Writer of the Past Teach the Technical Writer of Today?" *IRE Transactions on Engineering Writing and Speech,* EWS-4, 3 (1961), 69–76.

tone at the beginning of your message, since they often are intended to be accusatory.

Hitting upon the correct tone for your message means that you have to have a clear understanding of your purpose and your audience. Tone requires you to treat your readers as people. To avoid problems with tone, remember this variation on the golden rule: Write unto others as you would have them write unto you.

To show how poor tone dooms a message to failure, consider the following situation posed by Marvin K. Swift. Everyone in a given corporation, from porter on up to chief executive, is using the office copiers for personal matters. This minor piracy has been adding up in terms of cost and employee time. The general manager decides the time has come to lower the boom, and he does so with the following memo.[7]

```
TO:    All Employees
FROM:  Samuel Edwards, General Manager
SUBJ:  Abuse of Copiers

It has recently been brought to my attention that many of
the people who are employed by this company have taken
advantage of their positions by availing themselves of the
copiers. More specifically, these machines are being used
for other than company business.

Obviously, such practice is contrary to company policy and
must cease and desist immediately. I wish therefore to
inform all concerned--those who have abused policy or will
be abusing it--that their behavior cannot and will not be
tolerated. Accordingly, anyone in the future who is unable
to control himself will have his employment terminated.

If there are any questions about company policy, please
feel free to contact this office.
```

Edwards has, of course, ignored and insulted his audience. His tone is what gets him into trouble. (We'll return to this memo in the exercises at the end of the chapter.)

Pace

A final issue to consider in analyzing your audience is pace.[8] You are probably familiar with pace as a feature of oral communication. You've heard

[7] Marvin K. Swift, "Clear Writing Means Clear Thinking Means . . . ," *Harvard Business Review,* 51 (January–February, 1973), 59–62.

[8] The discussion that follows is adapted from a chapter on pace in Robert Rathbone's *Communicating Technical Information* (Reading, MA: Addison-Wesley, 1966), pp. 64–72.

speakers who lose their audiences or who put their audiences to sleep because their timing is off. Pace or timing is the rate at which you present information to your audience orally or on the printed page. Have you ever read a passage in a textbook and gotten lost in it? One reason for your difficulties with the passage may be its pace. Your mind lags behind your eye as your eye moves down the page.

In adapting your pace to your audience, you must worry about the complexity of your message. Information that is complex or difficult to digest needs to be written at a slower pace than information that is straightforward or familiar. One way of regulating pace is to look carefully at the ratio between the number of sentences and the number of ideas in each paragraph. With difficult material, there should be only one main idea in each sentence.

Carefully structured sentences can also help you adjust the pace of your writing. Make sure your reader can readily find the subject, the verb, and the object, if there is one, in each of your sentences.

You can further control pace by paying careful attention to punctuation, paragraph structure, word choice, format, and organization. Graphs and tables can provide ways for pacing quantitative data. Readers have difficulty digesting your data when your numbers end up crammed into sentences.

What do you make of the following memo sent from the building supervisor to the controller?

```
September 17, 1986

TO:       Mary Lee Chen
          Controller
FROM:     David Goff
          Building Supervisor
SUBJECT:  Painting Schedule, Industrial Center
```

Here's the information you requested about the schedule we have followed in painting various sections of the Industrial Center since we acquired the facility in 1965.

The foyer was painted in 1965, 1973, 1975, 1976, 1980, and 1982. The cafeteria was painted every other year since 1965. The exhibit hall was painted in 1965, 1967, 1975, 1977, 1978, and 1982. Odd-numbered offices on even-numbered floors were painted in 1965, 1968, 1971, 1974, 1977, 1980, and 1982. Even-numbered offices were painted in 1966, 1969, 1972, 1975, 1978, and 1982. The work areas that comprise the odd-numbered floors were painted as necessary. The men's rooms were painted in 1965, 1966, 1967, 1970, 1971, 1973, 1976, 1978, 1980, and 1982. The women's rooms were painted according to the same schedule. The lobbies outside each elevator were painted yearly until 1975, when wallpaper and paneling were installed.

> The fire towers, stairwells, and emergency corridors were painted every third year, beginning with 1965. All other corridors were painted every second year starting in 1966.

Were you able to understand all that data? No reader could. The problem lies with pace. The second paragraph is too long. It also contains too much quantitative data. The reader needs a simple table showing dates and areas painted to follow all the information the memo tries to provide:

PAINTING SCHEDULE[1]
Industrial Center

	'65	'66	'67	'68	'69	'70	'71	'72	'73	'74	'75	'76	'77	'78	'79	'80	'81	'82
Foyer	x							x			x	x				x		x
Cafeteria	x		x	x		x		x		x		x		x		x		
Exhibit Hall	x		x							x		x	x					x
Odd-numbered offices on even-numbered floors	x			x			x			x			x			x		x
Even-numbered offices		x			x			x			x			x				x
Restrooms	x	x	x			x	x		x			x		x		x		x
Lobbies	x	x	x	x	x	x	x	x	x	x	x^2							
Fire towers, stairwells, emergency corridors	x			x			x			x			x			x	x	
Other corridors		x		x		x		x		x		x		x		x		x

[1]Work areas that comprise odd number floors were painted as necessary.

[2]Wallpaper and paneling installed.

Robert Rathbone devised this simple guide to help writers determine the proper pace for individual prose units within a piece of technical writing. You can easily apply it to any letter, memo, or report you write.

Guide for Control of Pace

	Subject Area Unfamiliar?	Subject Complex?	Adjustment of Pace
Condition 1	yes	yes	Begin at slow pace; maintain slow pace.
Condition 2	yes	no	Begin at slow pace; accelerate to normal pace.
Condition 3	no	yes	Begin at normal pace; decelerate to slow pace.
Condition 4	no	no	Begin at rapid (or normal) pace; maintain rapid pace.

Source: Robert Rathbone, *Communicating Technical Information* (Reading, MA: Addison-Wesley, 1966), pp. 68–69.

The guide is based on the answers to two questions:

1. Is the reader unfamiliar with the general area of the subject?
2. Is the subject matter of the prose unit in question complex or detailed?

The phrase "normal pace" in column three refers to the pace most people use in their daily writing. Pace can vary throughout a longer document, but be careful not to lose your audience when you change your pace.

The following letter was sent by a major department store to all its charge customers. Initially, the letter's pace is right on target. What happens, however, after the second paragraph?

September 1, 1987

Dear Charge Account Customer:

A number of Credit Grantors have recently changed their policies to reflect changing conditions. We want to make absolutely sure that our Credit Policies are responsive to your needs and wants as our valued Customer and that they enable us to continue to extend Credit to our Customers.

This notification is being sent to advise you of two changes in your Charge Account Agreement: one relating to an increase in the rate of the FINANCE CHARGE and the other to a small increase in the MINIMUM MONTHLY PAYMENT on your account.

1. Under our new terms, any new FINANCE CHARGE in excess of 50¢ will be computed by applying a periodic rate of 1/2% per month (18% ANNUAL PERCENTAGE RATE) to the entire FINANCE CHARGE BASE. The effective date for this change will be December 1, 1987 thus applying to the FINANCE CHARGE BASE reflected on your December Charge Account Statement. Under our current terms, any new FINANCE CHARGE in excess of 50¢ is computed by applying a periodic rate of 1/2% per month (18% ANNUAL PERCENTAGE RATE) to the first $500 of FINANCE CHARGE BASE and of 1% per month (12% ANNUAL PERCENTAGE RATE) on any excess of FINANCE CHARGE BASE over $500.

2. Commencing with your December 1987 statement, the minimum monthly payment required for any New Balance up to $20.00 will be the entire New Balance. Any New Balance between $20.01 and $200.00 will require a minimum monthly payment of $20.00. This change represents a $5.00 increase in the MINIMUM MONTHLY PAYMENT schedule for New Balances between $15.01 and $150.00. Minimum Monthly Payments on higher balances will not be changed.

We wish to thank you for your continued patronage.

The first two paragraphs of this notice carry the reader along smoothly. The second two leave the reader choking on the writer's dust.

Readability

Readability, the efficiency with which your reader can comprehend everything you are trying to say, is closely related to pace. Technical writing texts have made much of the idea of readability, especially in terms of so-called readability formulas. These formulas attempt to gauge the complexity of written prose through easily applied mathematical analyses. Two of these formulas, Rudolf Flesch's Reading Ease Scale[9] and Robert Gunning's Fog Index,[10] have been especially influential and popular.

Their influence and popularity notwithstanding, readability formulas have little practical value in general in discussing writing as a process and in particular in analyzing an audience. If they have any value, readability formulas may be effective in judging the products of the writing process. They will tell you if you have written below, above, or at your reader's level of comprehension. They will not, however, tell you how to meet your reader's level of comprehension ahead of time or on a second try in a rewrite of your original message.

Readability formulas may also be suspect because the advice they offer writers is too simplistic and pat: use short sentences and short words. Earlier, when discussing the needs of an audience made up of laypeople or the general public, we did suggest that shorter sentences might better meet their needs as readers. But shorter sentences alone are no cure-all. Sentence structure needs to be considered too. A convoluted short sentence will not help any reader understand what you are trying to say. Word length is less important than word familiarity.[11]

While readability formulas may have limited value, readability is, nevertheless, an important concept if you wish to write a message comprehensible to your readers. Thomas N. Huckin, for instance, offers a sensible approach to readability.[12] While his essay focuses on readability and technical writing, its suggestions are applicable to any writing done in business and industry.

1. State your purpose explicitly, in such a way that your reader can anticipate the approach you are taking to the subject. Make it clear to the reader what issue or conflict you're addressing.
2. Make the topic of each section and paragraph *visually prominent*—by using

[9] Rudolf Flesch, *The Art of Readable Writing,* rev. ed. (New York: Harper & Row, 1974).

[10] Robert Gunning, *The Technique of Clear Writing,* rev. ed. (New York: McGraw-Hill, 1968).

[11] For a detailed discussion and reassessment of the value of readability formulas, see Jack Selzer, "What Constitutes a 'Readable' Technical Style?," pp. 71–89 in *New Essays in Technical and Scientific Communication: Research, Theory, Practice,* edited by Paul V. Anderson and others (Farmingdale, NY: Baywood, 1983).

[12] Thomas N. Huckin, "A Cognitive Approach to Readability," pp. 90–108 in *New Essays in Technical and Scientific Communication.*

headings and subheadings, and by placing topic sentences at the beginning of paragraphs.

3. Keep the topic in the reader's mind by referring to it frequently, preferably in the grammatical subject position of sentences.

4. Try to anticipate what reading style the reader is likely to use. If it's just skim-reading, concentrate on formatting, headings, topic sentences, visual aids, and other general guides to the informational hierarchy of the text. But if the reader will need to read for detailed comprehension or evaluation, employ techniques that help the reader in the step-by-step processing of information in short-term memory; the remaining guidelines describe some of these techniques.

5. Structure the text according to the nature of the information you want the reader to pay most attention to. If it's main ideas, use a hierarchical (general-to-particular) structure; if details, use a listing (coordinate) structure.

6. Once you've started referring to something by a particular name, *continue* referring to it that way. Don't vary your terminology just for the sake of variation.

7. When writing for nonspecialists, be sure to explicate the most important concepts in your text by using examples, operational definitions, analogies, or other forms of illustration. In other words, use familiar concepts to explain unfamiliar ones.

8. When writing for specialists, on the other hand, do not *over*explain. That is, do not use lengthy examples, operational definitions, analogies, and so on for concepts the reader is likely to already be familiar with. Instead, rely on the standard terminology of the field, even when such terminology is long and complicated.[13]

Purpose and Audience: A Few Parting Comments

In this chapter we have expanded our discussion of the PAFEO process by examining the issues of purpose and audience. In doing so, we have also made passing reference to format, evidence, and organization.

In the remaining chapters of Part I of this textbook, we will continue to expand our discussion of the PAFEO process. In Part II, we will apply that process to the various products you write or can expect to write in the world of work. In Part III, we will give you some "tools" to put the finishing touches on that product the PAFEO process has helped you to write.

Exercises

1. The Mobil Corporation regularly runs advertisements such as the one reprinted here. This particular advertisement originally appeared in *Time*. As the basis for class discussion, read the accompanying advertisement carefully. Then analyze the writer's purpose.

[13] Thomas N. Huckin, "A Cognitive Approach to Readability," p. 101.

A walk on the bright side

Four years ago the Carter administration came out with its highly publicized *Global 2000 Report*, warning that the world stood on the brink of doomsday because of overpopulation, pollution, and diminishing resources. The only hope, according to another government study, was more government action.

But a new look into the crystal ball by a group of independent scientists and academicians shows a far more hopeful future. Their conclusions have been published in *The Resourceful Earth*, edited by the late futurist Herman Kahn and Julian L. Simon, a professor of business and social science at the University of Maryland and senior fellow at the Heritage Foundation, the think tank which funded the study.

Some of the two dozen experts contributing to this anthology criticize the *Global 2000 Report* for being a political rather than a scientific document, and drawing conclusions not always supported by fact. They also fault the government report for basing its predictions on a few recent observations while ignoring major historical trends.

Examining many of the same issues as their government counterparts, these scientists foresee a less crowded, cleaner world with higher standards of living—provided the mind and imagination of human beings are not constrained by political or institutional forces.

The authors of the *Global 2000 Report* viewed rising population trends apprehensively. But the new study points out that the main cause of population increase is longer life expectancy—not only a sign of better agriculture and public health but a positive good in itself.

Moreover, the new study argues, population growth does not necessarily imply a more crowded life. In fact, the quality of life in the year 2000 should be superior, because improving living standards will produce better housing, roads, and transportation. Furthermore, the study argues that rising incomes will make pollution abatement more affordable—as is already happening in the richer countries of the world, where air and water quality continue to improve.

The new study also explodes the myth of increasing hunger. While acknowledging that many people in the world are still hungry, it points out that per capita food consumption is rising. And the world's food supply has been improving ever since World War II.

Another argument that pales in the light of fact: resource scarcity. The study shows that government predictions of ever more costly natural resources run counter to all historical trends. Because of technological advances and the development of substitutes, expenditures for raw materials as a proportion of total family budgets tend to drop over time.

While it would be foolish to pretend that the world's problems have been solved, mankind's prospects look a lot brighter than the doomsayers would lead us to believe. "The world is ready to turn its back on its pessimism, and is waiting to hear some good news." *The Resourceful Earth* says. We're pleased to share it with you.

Reproduced by permission of the Mobil Corporation.

- Would you describe the advertisement as purely informative?
- If you were the writer, how would you have stated your purpose?
- Considering the readership of *Time,* what might the writer assume about the audience?
- What does the format of the advertisement contribute to its purpose?
- What evidence does the writer present in support of the advertisement's purpose?

To expand your discussion of this advertisement, analyze it completely using the PAFEO process.

2. Even famous writers at times have to write for an audience in the world of business and industry. Samuel Clemens (better known to most readers as Mark Twain) fired off the following two letters, both notable for their tone. Discuss the advantages and the disadvantages of the tone Clemens uses in light of his audience.[14]

To Frank A. Nichols, Secretary,
Concord Free Trade Club

Hartford, March 1885
Dear Sir:

I am in receipt of your favor of the 24th inst., conveying
the gratifying intelligence that I have been made an
honorary member of the Free Trade Club of Concord,
Massachusetts, and I desire to express to the Club,
through you, my grateful sense of the high compliment thus
paid to me.

It does look as if Massachusetts were in a fair way to
embarrass me with kindnesses this year. In the first place
a Massachusetts Judge has just decided in open court that
a Boston publisher may sell not only his own property in a
free and unfettered way, but may also as freely sell
property which does not belong to him but to me—property
which he has not bought and which I have not sold. Under
this ruling I am now advertising that judge's homestead
for sale; and if I make as good a sum out of it as I
expect I shall go on and sell the rest of his property.

In the next place, a committee of the public library of
your town has condemned and excommunicated my last book
[*Adventures of Huckleberry Finn*], and doubled its sale.
This generous action of theirs must necessarily benefit me
in one or two additional ways. For instance, it will deter

---

[14] Quoted by John S. Fielden, "What Do You Mean You Don't Like My Style?" *Harvard Business Review,* 60 (May–June, 1982), 138.

other libraries from buying the book and you are doubtless
aware that one book in a public library prevents the sale
of a sure ten and a possible hundred of its mates. And
secondly it will cause the purchasers of the book to read
it, out of curiosity, instead of merely intending to do so
after the usual way of the world and library committees;
and then they will discover, to my great advantage and
their own indignant disappointment, that there is nothing
objectionable in the book, after all.

And finally, the Free Trade Club of Concord comes forward
and adds to the splendid burden of obligations already
conferred upon me by the Commonwealth of Massachusetts, an
honorary membership which is more worth than all the rest
since it endorses me as worthy to associate with certain
gentlemen whom even the moral icebergs of the Concord
library committee are bound to respect.

May the great Commonwealth of Massachusetts endure
forever, is the heartfelt prayer of one who, long a
recipient of her mere general good will, is proud to
realize that he is at last become her pet. . . .

Your obliged servant
S. L. Clemens

To the gas company

Hartford, February 1, 1891
Dear Sirs:

Some day you will move me almost to the verge of
irritation by your chuckle-headed Goddamned fashion of
shutting your Goddamned gas off without giving any notice
to your Goddamned parishioners. Several times you have
come within an ace of smothering half of this household in
their beds and blowing up the other half by this idiotic,
not to say criminal, custom of yours. And it has happened
again to-day. Haven't you a telephone?

Ys
S. L. Clemens

3. The following writing samples all have problems with purpose and audience.
Discuss the purpose and audience for each and the problems the writers face in
adapting their purposes to meet the needs of their audiences.

   Then, if your instructor directs you to, rewrite these samples so that they better
express their purpose and meet their audience's needs.

SAMPLE A

METROPOLITAN   INTEROFFICE MEMO
HOSPITAL

DATE: November 15, 1987

TO: Dr. Maria Cruz-Saenz, Neurosurgery
FROM: Alan McLane, Director of Medical Records
SUBJECT: Medical Record Dictation

The medical secretarial staff in medical records has
great difficulty in transcribing your dictation.

Recently you requested that one of the more experienced
secretaries be responsible for transcribing your work.
This is not possible.

The dictation equipment feeds your dictation to the
channel with the least amount of dictation. The
secretaries are each assigned different channels and are
responsible for transcribing all of the material on that
channel. Therefore, it is not feasible, from a logistical
standpoint, to have one secretary responsible for your
work.

All of the medical secretaries are familiar with
neurosurgical terminology.

It seems, however, that you speak too softly and too
quickly for them to understand what you are saying.
Perhaps if you speak somewhat louder and more slowly, the
problem you have encountered of having to fill in portions
of or correct segments of your dictation will resolve
itself.

SAMPLE B

(The fine print from the back of a state instant lottery ticket. Players rub off
boxes on the front of the ticket with a coin to see if and what they have won.
This particular instant lottery revolves around a baseball game in which "they"
have a score the player must beat. The player has nine chances, innings, in which
to run up a higher score than "they" do.)

Tickets void (and Lottery will not be obliged to pay) if stolen, unissued, unreadable,
mutilated, altered, counterfeit in whole or in part, miscut, misregistered, multi-
printed, defective, printed or produced in error, or blank or partially blank; if
anything other than 1 and only 1 digit (from 0 to 6) with matching caption appears
in each of 9 inning spots in gray ink; if anything other than 1 and only 1 digit
(from 1 to 9) with matching caption and legend of "Their" appears in "Their

Score" spot in gray ink; if anything other than 1 and only 1 prize amount (TICKET, $2, $5, $25, $500, $5,000 or $50,000) with matching caption appears in "Prize" spot in gray ink; if display printing irregular; if apparent or inserted Lottery symbols not confirmed by validation number; or if ticket fails any of Lottery's other validation tests. Liability for void or nonconforming tickets, if any, limited to replacement of ticket with unplayed ticket. Not responsible for lost or stolen tickets. Lifetime prize starts only at age 18 or older. Only highest prize paid per ticket. All tickets, transactions, and winners subject to Lottery regulations.

SAMPLE C

CITY & SUBURBAN   MESSAGE TO RIDERS
BUS & SUBWAY   by
AUTHORITY   Thomas P. Olson
                                                       General Manager

I don't like graffiti. Graffiti is vandalism. It's destruction of public property and it makes me angry to see it.

Over the years, we've tried many approaches to get rid of it and, at times, we were convinced we were winning the battle. There is one thing, however, we've learned about graffiti scrawlers. They are persistent.

But so are we.

We all know that marking up public property is a crime. More than defacing property we all share, it is a constant reminder that some people have no respect for the rights of others--your rights. It reminds us of danger. It says vandals work at will destroying our environment.

Graffiti on the transit system results in fewer riders. Some riders give up on CSBSA entirely because they don't want to be in a place that offends and frightens them. And that costs us money--lost fares.

In fact, if vandalism drives away only one out of every 100 CSBSA riders--just one percent--we estimate the total annual loss of revenue to be $1.2 million.

Then, if you add lost revenue to the cost of cleaning up the system and policing it, the total annual cost is well over $2 million.

Some people seem to think that graffiti and other acts of vandalism cannot be avoided. They throw their hands up and accept destruction as an inevitable part of our urban way of life.

CSBSA has not--and will not-- surrender to an army of vandals who want to slash, smash, and spray away our transit system. To combat graffiti we have devised a strategy that includes both internal cleaning and security measures and external legislative tactics.

Our heavy-duty cleaning crews are working constantly to remove graffiti or to paint over it.

The CSBSA Police Department has been enlarged to almost 80 officers. These men and women, working in uniformed and plain-clothed patrols, were trained at the Municipal Police Academy and have exactly the same powers as Metropolitan police officers. Our officers have arrested as many as 30 graffiti vandals in a single week, and, with your help, we intend to catch many more.

We have worked closely with elected officials, with law enforcement agencies and with the courts. The people we've caught are being made to pay for what they've done. In addition, the parents of young vandals are being held liable for up to $300 worth of damage caused by their children.

We need your help too. If you are angry about graffiti and vandalism--and what it costs you in transit fares and tax dollars--you can do something about it right now.

When you see the advertisements in your CSBSA vehicles that say, ""WE'RE GETTING A HANDLE ON THE VANDAL,'' remember the word HANDLE. When you add an ""S,'' it becomes a phone number: H-A-N-D-L-E-S. Call that number any time, day or night, if you see incidents of graffiti and vandalism. Your reports will permit us to get police officers to the scene, or to stake out places where incidents occur frequently.

All it takes is a phone call--and you don't even have to give your name. Just tell us where and when the incident took place, and whatever you can about the vandals.

Our region needs quality public transit--and CSBSA needs you. To help improve the system, please keep these phone numbers handy:

To Report an Incident of Vandalism

H-A-N-D-L-E-S

To Report Other Transit Security Problems

777-1234

To Stop a Crime or Save a Life

911

SAMPLE D

TO OUR VALUED CUSTOMERS

As you know, it has been our policy over the years not
to increase your fixed monthly cost for newspaper
delivery. Our costs for the different papers we distribute
have risen sharply, our labor rates have gone up
dramatically, and of course all fuels and related
byproducts remained high.

We have analyzed last year's profits and losses and the
economic forecasts for this year as applicable to our
business. We find that due to our increase in business
volume, experienced personnel, an increased profit margin
on some papers, and contemporary equipment, we once again
can maintain your fixed monthly cost for newspaper service
at the same rate it has been for years.

In past years, we have asked you at some time during the
year to voluntarily remit to us a sum that would cover our
spiraling expenses and enable us to operate a solvent
business with no increase to you on your bimonthly
statement. This system has proven to be 92 percent
effective and allows the less fortunate to renumerate an
amount they can afford.

The good news for 1986 is that we are asking you
voluntarily to remit the sum of $12.00 at a convenient
time of your choice--sometime in the next few months. This
figure reflects a 20 percent reduction to you from 1985--
as we realize a greater profit from the publisher and an
expected slower inflation rate.

With your cooperation we look forward to serving you at
a reasonable rate the best buy in America today--the home-
delivered newspaper.

Thank you,

SUBURBAN MORNING
NEWSPAPER DELIVERY
SERVICE

SAMPLE E

January 20, 1986

TO: All Medical Staff Members
FROM: Sarah Grotowski, Associate Director
SUBJECT: Improper and Profane Use of Private Dictating
          Number (555-9876)

Dr. Ethel McCormick announced at the annual staff meeting
on January 10, 1986, the preponderance of crank calls,
generally obscene and scatological, that we are receiving
through the direct dial number, 555-9876, to our record
transcription system.

I strongly suspect that some family members, visitors,
etc., who have access to your home or office telephones,
have also learned of this number and are using it for "fun
and games." The fun and games are not appreciated, since
the time of our transcriptionists has to be needlessly
used to screen through this "garbage," when their
productive time is so desperately needed to complete the
necessary dictating needs of each of you.

If this had just happened on an occasional basis, I might
have overlooked the situation. However, between December
14 and January 17, there were a total of 47 tapes with
these "crank" calls. There are profanities and obscenities
included on most of these that would cause a mule skinner
to blanch and which certainly reflect unpleasantly on
whoever is doing this "job" on us.

I urge you immediately to monitor the use of this direct
line from your home or office phone and to curb any
unauthorized use of it, if such use does in fact exist.

In the interim, we are securing a new number for you to
use. Effective, Friday, January 22, the new number will be
555-1234.

I am taking this action in hopes that the problem will be
resolved. However, if changing the telephone number does
not correct this situation, I will proceed with further
necessary action to track the telephone calls, and, if
necessary, prosecute.

4. Several months ago, a student from an out-of-state college was tragically murdered
   in a fast-food restaurant about five miles from your campus. The murder was
   widely publicized throughout the state.

You work in your college or university's Admissions Office, and you are given the following letter to answer. The original letter was simply addressed to "the Admissions Office." In responding, feel free to make up any school publications and send them, as well as any real school publications, along with your response. In responding, you do not really need the actual publications. Simply refer to them in the text of your letter.

Dear Admissions Office:

I would like some information about your school. My daughter has a girl friend who is trying to talk her into coming to your school. I am not against it, but I do not know very much about your school. I hope you can give me some information that will help me decide whether your school would be good for my daughter.

I read in the local paper that a boy was killed near your school. As a mother, I am concerned for my daughter's safety. Since Betty would be living on campus, I would like to know more about what that is like. I know that colleges and universities nowadays have what they call coed dormitories. I am not exactly sure what that means.

I know that I must sound like a terrible hick to you, but we live a good distance from your campus, and I know Betty would not be able to come home every weekend. I would like her to be living in a good atmosphere, one that would give a nice social life and not be dangerous.

I would appreciate any information you could give me about your school.

Sincerely,

Elizabeth McCorsky

(Mrs.) Elizabeth McCorsky

5. Return to Samuel Edwards' memo that we discussed earlier in this chapter (see page 43). You may remember that Edwards was trying to put an end to a minor piracy within his company. Everyone in the company from porter to chief executive officer was using the office copiers for personal duplicating.

   After rethinking the entire situation and Edwards' inadequate solution carefully, write a memo using the PAFEO process to solve the problem without creating new problems with personnel. There is no one correct solution here, but your solution should be workable without being overly complicated or bureaucratic.

CHAPTER 3

Evidence from Experience, Observation, and Research

When you have to write, you usually have to write in a hurry. Writing is, for most business people, just a part of their daily job responsibilities. Yet they often resent the task because it seems to take longer than it should. Surely you yourself have experienced difficulty in getting started on a writing project and keeping your ideas flowing smoothly to a conclusion. Nobody— not even professional writers—finds it easy to get started writing. Why is this the case?

Perhaps it is because we sense that writing is a complex task that requires many different mental abilities. To write, we first have to be able to generate ideas and put these ideas into language. Then we have to search out the relationships between these ideas and arrange them in some reasonable sequence. To make these ideas clear to others, we must find details and examples that illustrate our ideas in concrete fashion. It all seems overwhelming! So what do we do?

From personal experience, we as writers know all too well. It's time for a confession of past sins. Do any of the following practices sound familiar?

1. We procrastinate. We clean the desk, maybe even the room if we're scared enough of the writing. We sharpen pencils, change typewriter ribbons, eat, drink, visit—anything to avoid facing that blank, accusing look on the paper before us.

2. We plunge right in, hoping for the best. That is, we hope that inspiration will arrive if we struggle to stay afloat long enough, and that this inspiration will be strong enough to carry us on one perfect wave to the sands of success. While waiting for inspiration to arrive, we flounder along writing throat-clearing sentences like "It has come to our attention that there is some dissatisfaction in the area of customer relations."

3. Having written a sentence like the previous one, we start playing with the wording, looking for a way to make the sentence sound better. "We have been informed that there have been complaints. . . ." Or "Our representative, Mr. Stephen Markku, has called to our attention. . . ." Meanwhile, time is passing, and anxiety is rising.

4. We try to do all of the complex parts of the writing process simultaneously—not just simultaneously, but perfectly! We try to think up ideas, find words that express them, put the right words in the right order, arrange the ideas, and establish the right tone. Since we're really trying to arrive at the product without recognizing the complexity or necessity of the writing process, we usually find ourselves stuck. Either we are unable to go on, or else we inch our way through a torturous first paragraph, spending hours or days trying to get it right. We make life hard for ourselves by editing our writing before even giving the ideas a chance to emerge fully.

Have we correctly diagnosed your approach in one of these descriptions? More than once, we have found ourselves bogged down in one of these self-defeating approaches.

There *is* a better way. That is the good news to be developed in this chapter. There are powerful and effective strategies that have been used, consciously or unconsciously, by writers who recognized that writing was too complex when tackled all at once. These writers had to find ways of breaking down the process into manageable parts. You can do the same thing to make your writing easier and better.

In the following pages, we'll introduce you to some systematic techniques you can use to help you find something to say by probing your memory, your observations, and your research. You can use these techniques to generate the ideas and collect the information you will use as evidence to support your purpose.

To see how these techniques can work, let's consider a writing situation that we introduced you to as Exercise 4 at the end of the previous chapter.

The McCorsky Case

Your first job out of college happens to be with the Admissions Office of an urban university. The Director of Admissions drops the following letter on your desk with a note saying, "I think we ought to respond with a personal letter in addition to any brochures we send. See what you can work up. Let me see it before you send it. Thanks."

The background is this: Several months ago a student from an out-of-state college was tragically murdered in a fast-food restaurant about five miles from your campus. The murder was widely publicized throughout the state. Now, here's the letter to which you must respond.

Dear Admissions Office:

    I would like some information about your school. My daughter has a girl friend who is trying to talk her into coming to your school. I am not against it, but I do not know very much about your school. I hope you can give me some information that will help us decide whether your school would be good for my daughter.

    I read in the local paper that a boy was killed near your school. As a mother, I am concerned for my daughter's safety. Since Betty will be living on campus, I would like to know more about what that is like. I know that colleges and universities nowadays have what they call coed dormitories. I am not exactly sure what that means.

    I know I must sound like a terrible hick to you, but we live a good distance from your campus and I know Betty would not be able to come home every weekend. I would like her to be living in a good atmosphere, one that would give her a nice social life and not be dangerous.

    I would appreciate any information you could give me about your school.

Sincerely,

Elizabeth McCorsky

(Mrs.) Elizabeth McCorsky

It doesn't take much analysis to see that there is quite a bit at stake in how you answer the letter. If you send out something that does not respond to the concerns, both expressed and implied, of this parent, there is a good chance that the student will choose another institution. You may want to enclose some informational brochures with your response, but the way you handle the content and tone of your letter is likely to be a decisive influence on the way this woman will feel about her daughter's attending your college. Where do you begin?

Remember: If you try to do everything at once, you are lost. Let's see how the use of the PAFEO process could be a helpful starting point.

Purpose

As you may remember, we suggested that you try to put your statement of purpose in one sentence and that you state as exactly as possible what you want the reader to do or think as a result of reading what you have written. The model for your purpose statement might look something like this:

My purpose is to _____
so that Mrs. McCorsky _____.

The second blank reminds you that you are writing to get Mrs. McCorsky to change her thinking. To do that, you have to have some idea of what that thinking is. Her letter tells you both explicitly and implicitly what problems she wants you to solve for her. Although she says she wants general information about the university, her deeper concerns are apparent throughout her letter. What do you see as the number one priority you must address in your reply? If you said safety or security, we agree. Mrs. McCorsky mentions the murder as one cause of her fear, but she also expresses uncertainty about the moral climate of coeducational dormitories. The word "dangerous" reappears in the last paragraph. Therefore, what she really needs from you is information that will reassure her so she will not be opposed to her daughter's attending your institution. You may also want to win her over as a positive supporter who will encourage her daughter to choose your school. Thus your purpose statement might look something like this:

My purpose is to clarify questions concerning students' safety and the general atmosphere of residence halls so that Mrs. McCorsky will be persuaded that the university provides the quality of life she wants for her daughter and will encourage her to attend.

Now you're off to a good start. You know what you want this piece of writing to do for your audience. You have a goal, and that goal will shape your choices as you proceed through the writing.

In real-world writing, remember, you are often plagued by the problems of multiple, sometimes conflicting, purposes to be served for several audiences, each one of which has different needs. In this assignment, for example, you can't forget that your supervisor, the Director of Admissions, is also an audience, and that one of your secondary purposes is to persuade her of your ability to confront and solve problems.

Thinking carefully about purpose will help you establish priorities when you have multiple purposes and audiences. You can develop a list of problems the writing must solve for each audience. As you think about how you can achieve your purpose with a particular audience, remember what you learned in the previous chapter about analyzing your audience. Let's see how such analysis would apply to Mrs. McCorsky.

Audience

If we apply the appropriate items from the checklists in Chapter 2, the following picture of Mrs. McCorsky might emerge.

Since Betty is apparently her first child to enter college, Mrs. McCorsky is probably in her early forties. She lives far enough away from the campus that she has some fear of the area. Her letter suggests that she herself has not attended college and that she feels somewhat uncertain and even intimidated by the world her child is now entering. Her fear that she may seem to be a "hick" tells us much about her insecurity in facing this new situation.

What does all this information mean to you as you sit down to write to her? It means you must not bury her in details or in academic terminology she does not care about and will not understand. It means you must avoid anything that sounds as if you might be talking down to her. It means you cannot take any background knowledge of college life for granted. Finally, it means you are trying to communicate with another human being who has taken the risk of exposing her own fears and uncertainties because she cares so much about her daughter's future. Mrs. McCorsky deserves to be dealt with in human rather than in bureaucratic terms.

Format

The format will obviously be a letter in this instance, but what kind of letter? Long or short? Personal in tone or official? Even though her request is a short letter, it deserves a full reply that may run more than one page. Addressing the underlying concerns in her request is not a matter to be handled in a businesslike short reply; the personal tone of the reply is also important.

Evidence

Some of the evidence you will want to use to back up your statements about the safe atmosphere of the university will be in the form of printed publications you will enclose. You will probably look for those with the finest pictorial presentation, depending on the pictures to convey some of your message about the atmosphere of the institution.

What kind of evidence should be mentioned in the letter itself? Perhaps a quotation from a student would be helpful, especially if that student were from Betty's hometown. You might want some information on the adequacy of the Campus Security Force. One pitfall, which several of our students' responses have made us aware of, is to overreact with so many security statistics that the campus sounds like an armed camp.

Organization

Get to the most important points first. Since Mrs. McCorsky is primarily concerned with her daughter's safety, respond to that need as soon as possible. Then you can work your way down to less important concerns, taking the opportunity to make some points on your own that will help persuade Mrs. McCorsky that your school offers the best learning climate for her daughter.

Getting Started: Invention Strategies

By doing a careful analysis of purpose and audience, you take the first important steps in the process of finding, developing, and selecting the content you will use to inform or persuade your audience. In classical rhetoric, this process was known as *invention*. Invention strategies are useful in generating the ideas and facts that you will select and organize to achieve your purpose with a particular audience. Not every strategy appeals to every writer, nor is every one suited to every writing situation. For those reasons, writers need to have a range of possible strategies in their arsenals. Here are some of the strategies experienced writers use in generating possible evidence to be used.

Strategy 1: Freewriting

The main idea of this strategy is to forget your need to edit or correct everything and just go ahead. Write continuously for a predetermined time, say, five or ten minutes. Don't pause, don't change anything that doesn't sound right, don't even stop to search for the next word; if it doesn't come immediately, keep repeating the previous word until it does.

If you've never tried this technique, you may be skeptical about whether anything good can come of it. Peter Elbow, one of the strongest advocates of this technique and a teacher-writer of great creativity, claims the following advantages for the technique:

1. It helps you learn to separate the producing process from the revising process.
2. It helps you think of topics to write about.
3. It helps you write without thinking about writing, naturally and powerfully instead of hesitantly and cautiously.[1]

After the allotted time, read what you have written. Read curiously, not critically. You want to look for ideas that seem to be pushing you to write about them. Don't censor yourself or look for hidden traits of character. Freewriting is supposed to be an exercise in pump-priming, not one in critical analysis.

[1] Peter Elbow, *Writing with Power* (New York: Oxford, 1981), pp. 13 ff.

If you're not held up by writer's block but instead have a pretty clear idea of your goal and a grasp of the problem you're facing, another strategy might be more appropriate. For instance, you know you have to persuade Mrs. McCorsky that her daughter will be reasonably safe and happy on campus so that the good woman will approve of her daughter's attending your school. What should you consider in your letter? Try Strategy 2.

Strategy 2: Brainstorming

Brainstorming is a more goal-oriented process than freewriting. You are trying to come up with ideas related to a particular problem. Otherwise, brainstorming requires some of the same rules as freewriting: no censoring, no editing or polishing. You don't have to set a time limit, but some writers feel time pressure helps. Jot down words or phrases as they come to mind. Try to capture anything you can think of that's relevant to the problem. If you get a word or a whole paragraph or even a sketchy diagram of a few ideas, the technique is working for you.

Let's see what happens if we apply this technique to our reply letter. Here's a possible list produced by brainstorming.

1. Pleasant campus. Send her pictures?
2. Security force large enough and functions well.
3. Dorm students not particularly fearful, according to evaluations.
4. Murdered student took unnecessary risks in bad neighborhood (not ours) at very late hour. Another school—Why relevant?
5. Students here taught rules for greater safety in today's world.
6. Escort system for students studying late at library and returning to dorms.
7. School has good relations with neighborhood. No guarantees.
8. Coed dorms mean that men live on one floor, women on another, and do not share baths or beds.
9. Coed dorms are supposed to help students develop individual responsibility and healthy social life.
10. Impossible to guarantee safety in today's world—on this campus or anywhere else.
11. A visit? Best if Mrs. M. and Betty see for themselves. When would be a good time?
12. Social life—dorm parties, sports, concerts, nearby restaurants, sports teams, orchestras, movies, theatres, campus events.
13. Varied life of the campus community itself may be a positive experience for someone whose previous experience has been in another town or city.
14. Try to understand how I would feel as a parent. Note Betty is fortunate in having a friend from her hometown who is also interested in attending.

To make the most effective use of your brainstorming list, however, you must first do a good job of clarifying your purpose. If you have a clear goal, you can shape your list accordingly.

Let's look at the list with our specific purpose statement to guide us:

My purpose is to clarify questions concerning students' safety and the general atmosphere of residence halls so that Mrs. McCorsky will be persuaded that the college or university provides the quality of life she wants for her daughter and will encourage her to attend.

Let's remember also the needs of our primary and secondary audiences, Mrs. McCorsky and the Admissions Director. Mrs. McCorsky needs clarification and reassurance; your boss needs evidence that you have done a persuasive selling job in recruiting this student. Now back to the list to see if it needs a little rearranging, addition, or subtraction.

Keep two principles in mind:

1. Your message must be reader-based. The most important words you as a writer can use are "you" and "your."
2. If your audience is unfamiliar with your subject, stress the familiar, either by using an analogy, or by defining the unfamiliar.

With these principles in mind, perhaps we should place our last item first and try to add something stressing the familiar or defining the unfamiliar.

If we group the other items around Mrs. McCorsky's two principal concerns—the safety of the neighborhood and the atmosphere of the residence life—we might come up with a rearrangement something like this:

14. Try to understand how I would feel as a parent. Note Betty is fortunate in having friend from her hometown who is also interested in attending.
10. Impossible to guarantee safety in today's world—on this campus or anywhere else.
3. Dorm students not particularly fearful, according to evaluations.
7. School has good relations with neighborhood. No guarantee.
2. Security Force large enough and functions well.
6. Escort system for students studying late at library and returning to dorms.
5. Students here taught rules for greater safety in today's world.
13. Varied life of the campus community itself may be a positive experience for someone whose previous experience has been in another town or city.
8. Coed dorms mean that men live on one floor, women on another, but do not share baths or beds.
9. Coed dorms successful in developing individual responsibility and improved social life.
1. Pleasant campus. Send her pictures?
11. A visit? Best if Mrs. M. and Betty see for themselves. When would be a good time?

Note that we dropped item 4 as irrelevant to our purpose and the audience's needs. (That item reads, "Murdered student took unnecessary risks in bad neighborhood—not ours—at very late hour. Another school—Why relevant?) The murder is a symptom of Mrs. McCorsky's concern, but it is not related to your campus's security.

If you have written your letter to Mrs. McCorsky as one of the writing assignments at the end of Chapter 2, you might want to compare your reply with the list above. Have you touched upon the same items? Have you found others that you think ought to be included? Can you justify them as necessary to your purpose or to the needs of your audience? Remember, there is no single reply that is right or wrong; the letter you have written can be evaluated as to how well it meets the concerns represented in PAFEO. Later in the chapter, we will present one writer's effort to answer Mrs. McCorsky according to the PAFEO process.

Keep in mind the following points whenever you use brainstorming: It is most effective if you can make it goal-directed by clarifying your purpose and analyzing your audience before you begin. After that, be as playful with ideas as you can. Don't worry about phrasing them right; just get them down. You can polish them or junk them later.

Strategy 3: Probing Questions

Another way to get going is to ask yourself the right questions. The journalist's traditional questions may be useful to you in certain kinds of business writing, especially that which is mainly informative.

The newswriter begins a story by trying to answer as many of these questions as possible: *What, who, why, when, where, and how.* Could these questions help you generate some ideas for your reply letter?

What?	Safety.
Who?	Betty, a young woman from a smaller town, and her mother.
Why?	Campus location; murder publicity aggravates fears, especially in parents from distant areas.
When?	All times, but especially at night.
Where?	Neighborhood and on campus, especially in dormitories. Concern for physical safety and for moral safety in context of coed dorms.
How?	Security force, escort service, checkpoints at entrances, men on first and second floors, women on third floor. Cooperative efforts with neighborhood associations.

The more fully you can answer the questions, the better your focus on the problem. The questions are a quick way of zeroing in on some of the main ideas you will want to include in your letter to Mrs. McCorsky.

Strategy 4: Role-Playing

Imagine that Mrs. McCorsky and Betty have just walked into your office and expressed the ideas in her letter. What would you say to them, knowing you only had a few minutes to get your point across? More importantly, what questions would she and Betty have? What answers could you offer?

This technique works well in helping you to get ideas and to put them into words. Talk to your audience, whom you now see more clearly as two audiences—the mother and the daughter. Understand their concerns by playing their roles.

If you wish, act out the roles using a tape recorder. When you play the tape, take notes on the points you want to include. You will have plenty of material to draw on when you actually sit down to write your letter. In this instance, the presence of the two different audiences will make you aware of the problem of meeting the different and sometimes conflicting needs and expectations of mixed audiences reading the same document. You will have to choose your primary audience.

Strategy 5: Incubation

You may like our last invention strategy best of all: put away the writing problem you are working on and just forget it for a while. Providing you have focused sufficiently on the problem you're trying to solve, you may find that it seems to have solved itself in the hour or the day in which you have left it alone.

Time after time, teachers have probably suggested that you allow yourself enough time on an assignment so you can read it over with a fresh eye the next day. This approach often works surprisingly well. If you have only three hours to spend on a writing task, spend it in several chunks, leaving incubation time between each chunk. Writing is like a baloney sandwich; leave in in the drawer for a day or two, and you will smell it if it starts to go bad.

If you have really focused on your problem, your mind will be working on it even when you don't realize you are working on the idea. When you return to the writing, you may find that the idea you need comes to mind when you least expect it, and the whole writing problem begins to seem easier. Don't take a chance on losing the idea that emerges from the shell in the incubator; make a quick note of it before you lose the fragile creature.

The Product of Invention Strategies: A Reply to Mrs. McCorsky

As a result of applying the invention strategies discussed, one of our students came up with the following lengthy reply to Mrs. McCorsky. You might want to compare it with your own and discuss its effectiveness with your classmates.

```
Dear Mrs. McCorsky:

    Your concern for your daughter's safety at college is
certainly understandable and admirable. I'm sure I will
feel the same way when my own daughter leaves home for the
first time. Although no one can offer absolute guarantees
about safety in today's world, I hope I can reassure you
```

that we too are concerned about each student and that we do everything we can to make living at this university a safe and satisfying experience.

Our present students seem to be happy with our efforts, according to the evaluations they complete each year. I think it would be useful for Betty to talk to some of them, and I will be glad to arrange a meeting if you would like. We have a junior named Megan Sands who attended the same high school Betty did. She tells me she would be glad to talk to Betty; she thinks she understands exactly how Betty must feel. Megan has been quite pleased with her residence experience here.

Reports of violent events near a campus naturally raise parents' fears. We believe that a constructive policy of maintaining friendly and cooperative interaction between the university and its neighborhood is one of the best ways of building a more secure atmosphere, and we work hard at maintaining communication between the university and its neighbors. At the same time, we recognize our responsibility to provide a trained, adequate security force that is on duty 24 hours a day. The Security Department attempts to foster in our students good habits that will provide them with the best protection for life in today's world.

To give one example, students are urged to make use of Security's escort service. If your daughter happens to be studying late at the library at night, she need only dial a number to obtain a security escort back to her residence hall. Although you always read more bad news than good in the press, we would like Betty to discover what this campus community has to offer as an exciting place of discovery and social and cultural opportunities.

Naturally you want Betty to enjoy living on campus and to develop maturity as a result of living away from home. Coeducational residence halls can help develop the responsibility and maturity of students. In our halls, men students occupy the first floor and women students the second or third floors. We have found that the occupants of a floor quickly become a "family," sharing concerns and friendship. The residents of a particular hall develop a sense of community under the leadership of resident assistants, junior or senior students chosen for their leadership ability.

I am enclosing some illustrated brochures to give you a pictorial sense of the campus, but I think the best way of answering any questions you may still have would be for you and your daughter to visit the campus for a day. If you will telephone me at the number on this letterhead, I will be delighted to arrange a time for such a visit. If

you are as pleased as I think you will be with the campus
and the people here, I am sure Betty will want to confirm
her application before she leaves the campus.

Sincerely,

Michael Juhanni

Michael Juhanni
Assistant Director, Admissions

Exploring Sources of Information

The invention strategies we have discussed will help you probe your own remembered or observed experiences. But you also need strategies for researching information in a systematic way.

Here are five important sources you may find useful in your efforts to gather material you will use as evidence to support your purpose. Each has some advantages and disadvantages, and we'll examine those briefly.

1. Personal observation or experience.
2. The testimony of others, collected through interviews, questionnaires, consultation.
3. The opinion of experts, found directly in books or journals or indirectly through computer-assisted research.
4. Statistical data from company, industry, or government publications.
5. Computer-assisted research systems.

Personal Observation

Your personal experience and observations add vividness and power to what you are saying—simply because they *are* personal. Nobody knows what you have experienced as well as you do. Personal examples and illustrations hold the reader's interest by their concreteness and specificity. They are powerful tools for helping your reader see what you mean. But to make the most of your personal observations, keep two cautions in mind: first, observe carefully, taking notes on what you see; second, be sure your experience is typical and representative before you generalize upon it.

Testimony of Others

The most direct way to get information from others is to ask them personally in an *interview*. That sounds simple enough, but it seldom is. There are many points to consider. Whom should you interview? Don't waste time by interviewing someone who can't tell you what you want to know. What should you ask? A good interview demands much homework. You ought to have an idea of the form you want the interview to take and also have a list of questions prepared in advance.

Another way of gathering information is to ask for expert opinion by letter. You are imposing on another person's time and attention when you write such an inquiry letter, so be brief and courteous. Whether you receive an answer or not may well depend on how well you write the letter.

When you have exhausted other means of collecting information and still do not have what you need, you may decide to use a *questionnaire.* It is not a decision you should arrive at lightly. Questionnaires are challenging even to experts. How do you make up reliable questions, ones that really measure what you intend? How do you ensure validity, meaning that the reader's interpretation of the question matches your own? And how do you select the type of question best suited to your purpose? Any of the preceding questions is enough to give one pause in designing a questionnaire.

But if you're convinced the questionnaire is the best route to the information you need, here are a few points to keep in mind:

- Clarify the purpose you wish the questionnaire to serve. Are you developing statistical information, measuring opinion changes, seeking an evaluation of a product or service? What purpose will the information you develop serve when you report it?
- Remember that you have two audiences: (1) those who will answer your questionnaire, and (2) those who will read the report you write summarizing your findings. To develop a good questionnaire, you need a clear idea of the needs of readers of your report.
- Determine the kind of information you need and frame the questions accordingly. Basically, questions may be either open-ended (essay) or closed (objective). Open-ended questions are introduced by such words as "discuss," "compare," "comment on," or "explain." Such questions are useful if carefully focused. They are not easily summarized in a report. Closed questions are more often used because the limited choices offered the respondent make them easier to tabulate. Whichever you use, be careful not to ask too many questions. Make each one count. Respondents tend to discard lengthy questionnaires.
- Include one idea per question.
- Select the format for questions that seems best for your purpose and stick to it throughout the questionnaire. Here are some examples of the most commonly used kinds of closed questions.

Multiple choice: You have just seen your favorite date with someone else after you had been told a heartbreaking story about studying for a test. What manner of murder would you prefer?
____ A. axe murder   ____ B. poisoned beer   ____ C. strangulation

Yes or No: Do you eat the Surprise Stew served in the dining hall?
____ Yes   ____ No

Fill-ins: How many gerbils do you now own? (If none, write "none.") _____

Ranking: How would you rank the cheeseburgers sold by each of these fast-food chains? Place a number in each space to indicate your order of perference.
____ McDonald's   ____ Burger King   ____ Wendy's
____ Other
(Write in name) _____

Rating: Place a check in the appropriate box to indicate the importance of each item to you.

	Very Impt.	*Somewhat Impt.*	*Not Impt.*
pizza			
fries			
burgers			
beer			
soda			
milk			
fruit			
veggies			

Expert Opinion

Aside from that solicited in inquiry letters, you will look for expert opinion in books or journals dealing with your topic. You will find a list of basic reference tools in Chapter 12.

Often the experts disagree. In such cases, we have to evaluate their arguments and determine their credibility. To take an example, an expert in the pay of a power company may not be as objective on the subject of nuclear energy as another scientist who is working independently at a university.

Library Research

You can get statistics and up-to-date information on almost any subject if you know how to use reference books. Facts and figures can be found in almanacs such as *The World Almanac and Book of Facts* (New York: Newspaper Enterprise Association, published annually) and in yearbooks such as the U. S. Bureau of the Census' *Statistical Abstract of the United States* (Washington, D. C.: Government Printing Office). As a business writer, you will want to familiarize yourself with Bernard S. Schlessinger, *The Basic Business Library: Core Resources* (Phoenix: Oryx Press, 1983). In addition, Eugene P. Sheehy, *Guide to Reference Books,* 9th ed. (Chicago: American Library Association, 1975; Supplement, 1980), is an invaluable guide to abstracts, books, indexes, and periodicals in each discipline. Your reference librarian can help you gain access to the rich informational resources of government documents. These items provide information on an enormous range of subjects. We will have more to say on the subject of library research in Chapter 12.

Note Taking and Card Sorting

There is a time-tested way of taking notes that will save you much labor in the long run because it will allow you to handle the information and ideas you have collected and generated with greater ease. In the beginning, you may think it is more convenient to write everything in your notebook. But thousands of researchers have found otherwise. Trust us. Try the card method. (See the illustrations on page 74.)

Use three-by-five-inch index cards for keeping track of the books and other sources of information consulted. Write the necessary bibliographical information (author, title, place of publication, publisher, and date of publication) from the card catalog or reference work. Don't rush. Nothing is more frustrating than to have to retrace your steps to find information that you should have written down the first time. Take time to write out the full information.

Even use four-by-six or five-by-eight-inch cards for your notes, especially if your handwriting is large. When you have a pile of cards, write a brief identifying label at the top of each card (like *Marketing* or *Production*). You can then easily sort the pile of cards into several categories, each bearing the same identifying label.

David Ewing suggests the following useful list for classifying information in his book, *Writing for Results.* You won't need all of them in a single writing task, but the list can help you identify the principal categories you want to use for a particular writing job.

- New findings, conclusions, and/or questions.
- Recommendations and proposals.
- Alternative plans, programs, sequences of steps.
- Problems and limitations.
- Methods, processes, techniques employed.
- Description of conditions and trends.
- Background.
- Discussions of special aspects or considerations.
- Stages of time, actions, process.
- Places where the operations, problems, are found.
- Levels and areas of concern.
- Degrees of significance.
- Definitions of terms.
- Arguments pro and con, conditions favorable or unfavorable.[2]

[2] David W. Ewing, *Writing for Results,* 2nd ed. (New York: Wiley, 1979), pp. 99–100.

Example A

> C.H. Knoblauch,
>   "Intentionality in the
>   Writing Process,"
>   College Composition and
>   Communication, 31 (May, 1980),
>   153-159.

Example B

> Many communications serve several conflicting
> purposes for different audiences. Ex: a letter
> may have to - inform customer of bad news
>   - explain reasons
>   - persuade customer of co.'s good will
>   - cover legal requirements
>   - satisfy the supervisor
>   - provide a record of the facts
>
> A problem-solving approach may help clarify
> the purposes so the writer can establish
> priorities. Writer would develop list of
> problems the communication must solve for
> each audience (p.158).

Once you have divided your pile of note cards into several smaller piles, you can arrange the cards in each pile according to the organizational pattern you have selected as most appropriate to your purpose, audience, and subject material.

If you can incorporate some or all of the techniques discussed in this chapter into your approach to a writing job, you need never again worry about how to get going. Instead of depending on old habits that don't work very well, like trial-and-error or a wish-and-a-prayer, you will have a systematic way of thinking about your subject that can help you with your first challenge—that of generating ideas about which to write.

In addition to the techniques you have learned for probing your own mental resources, you have been introduced to some other important informational sources and methods. Remember: No single strategy or technique fits every writing situation. Writing involves a continuing series of choices. To make effective choices, you need to have a range of possibilities from which to choose. Providing you with that range has been our aim in this chapter. If you consider what we have said, you will have some useful tools at hand. The more you practice using these tools, the more your skill with them will grow.

Exercises

1. Following is a list of campus-related topics. We will return to this list later in Chapter 12 when we discuss the formal report.[3] For present purposes, use the list as a basis for class discussions aimed at exploring the advantages and disadvantages of the invention strategies we introduced in this chapter. In your discussions, assume that each topic is in reality a problem on your campus.
   a. On-campus parking.
   b. Space assignment for student activities.
   c. Weekend or late night access to the library or food services.
   d. The justification for tuition increases.
   e. The effectiveness of admissions recruiting efforts.
   f. The effectiveness of campus public relations publications.
   g. Campus security.
   h. Discipline in the dormitories.
   i. Access to campus computer terminals.
   j. Academic and other support services on campus.
   k. Writing problem(s) on campus.
   l. A host of curricular issues.

2. Return to the list of topics provided in Exercise 1. As an additional exercise in classroom discussion, select one of these topics and design a questionnaire to sample student opinion on that topic.

3. Your college or university has just announced a 10 percent increase in tuition and fees. Students are understandably concerned; in some cases, they are outraged. As a responsible spokesperson, you have been chosen to meet with an appropriate administrator who has agreed to respond to presubmitted questions from students. The interview will last no more than thirty minutes. Draw up a list of questions that you would submit in advance to the administrator.

4. To acquaint or reacquaint yourself with your school's library, visit the reference collection to find out the answers to the following questions or information on the following topics. Remember to keep track of where you found the information or answers.
   a. Which of the fifty states has the largest land mass?

[3] This list originally appeared in Kevin J. Harty, "Campus Issues: A Source for Research in Business and Technical Writing Courses," *Teaching English in the Two-Year College,* 12 (October, 1985), 223.

    b. In which state are the largest number of corporations incorporated?

    c. Biographical information on the presidents of the two largest American automobile manufacturers.

    d. Which state has the largest corn harvest?

    e. Which metropolitan area includes the largest number of colleges and universities?

    f. Who is the second highest paid official in the federal government? What is that official's salary?

    g. Which nations belong to OPEC?

    h. Statistics by country on aluminum production worldwide in 1983.

    i. Books and articles written on the Hilton Hotel chain.

5. Using the *Business Periodical Index* and the *Wall Street Journal Index,* collect notes on file cards dealing with the subject of writing in the business world. Set up several categories and label your cards accordingly. The list of categories on page 73 may prove helpful.

6. Imagine that you are Megan Sands, the young woman we mentioned earlier in this chapter who is from Betty McCorsky's hometown. The Director of Admissions has asked you to write to Betty too, encouraging her to attend your college or university.

    You remember Betty from high school as an intelligent, socially aware young woman, unhappy with the many restrictions of her home town (and of her overly protective mother?).

    The Director of Admission will not see your letter, but Betty's mother might. Write a purpose statement for your letter to Betty. Then consider which of the invention strategies will work for you in this letter. After a period of some trial and error, write the letter.

CHAPTER 4

Organization and Format

In this chapter, we will examine some of the choices you have in arranging your ideas in an order that makes sense to both you and your reader. This establishing of a pattern is what we call *organization*. In the second part of this chapter, we'll have a look at how the message ought to appear on the page. You will see that this matter of choosing the most attractive and most effective *format* contributes greatly to the communication process.

Organizing for Your Audience and Yourself

If your invention strategies work the way they're supposed to and you generate some good ideas, these ideas will get you nowhere unless you can organize them. "For it is by organizing facts and ideas properly that the writer establishes meaningful relationships," said David W. Ewing, former editor of the *Harvard Business Review*, "and it is by seeing such relationships that the reader understands the material."[1]

[1] David W. Ewing, *Writing for Results*, 2nd ed. (New York: Wiley, 1979), p. 94.

So the organizing process helps both reader and writer. For the writer, organizing is part of the discovery process, the working out of the puzzle. (Which things fit together? How can I group these ideas into related packages?) For readers, organization is the current bearing them swiftly and easily along with the writer's thought. Lacking the comforting sense of direction provided by a clear arrangement, readers are riding the rapids and hitting every rock in sight. And after a rough ride, they experience the frustration of getting nowhere. Every piece of writing ought to give the reader a sense of starting somewhere and arriving somewhere else. As one of our students succinctly summarized this advice: "Writing should be a real trip."

The need for organization is so obvious that no one would question it. But the natural laziness of human nature and the pressures of time provide us with excellent reasons for neglecting the obvious.

"I didn't have time," says the harried writer.
"I can't follow this," says the even more harried reader. "It's all over the place."

There is no easy plan for organizing that is adaptable to every writing job you encounter. Where then is one to begin the process of choosing the best arrangement of ideas?

If you have thought about your purpose and audience and used invention strategies to generate ideas, you are already well on your way to organizing your material. Organizing is not a separate, independent process. The organizational pattern you select ought to result from the writing situation in which you find yourself, and that situation includes the facts and opinions you are presenting, your purpose in presenting them, and the needs of your audience.

Look at what you have. Read your purpose statement again. Remind yourself of the audience(s) who will read your material and the purposes to which they will put it. Now look at the ideas you have generated in your invention process. Have you raised questions requiring further research if you are to have the evidence you need? What additional facts or data should you have on hand before going further?

Before you try to organize your material, make sure that you have collected sufficient information to provide the evidence you need. One of the advantages of a systematic invention process is that it can help you see what you don't know; the questions you raise for yourself will point out the need for certain information that memory alone cannot supply. Unless you have adequate data, your writing can't achieve its purpose.

General Principles to Keep in Mind (Applied PAFEO)

1. Make sure that everything you write has a *beginning, middle,* and *end,* and that each segment fulfills its *purpose.*

- The *beginning* should tell the reader what you're going to do in this piece of writing and try to create some interest by explaining why this writing is needed.
- The *middle* presents specific *evidence* supporting your purpose or thesis.
- The *end* may summarize your findings, present conclusions, or make recommendations for action.

2. Analyze your reader's likely response to your materials and *organize* accordingly. If the reader appears to be neutral or favorable to your position, organize according to a *deductive* plan, putting the most important ideas first and the supporting ideas following them. If the reader is more likely to be resistant or resentful, however, lead up to your conclusion gradually in *inductive* order, presenting reasons and details before you present the conclusions or recommendations.

3. Put the ideas you want to emphasize at the beginning or the end (or in both places). Create a *format* that will allow the reader to see the points of emphasis easily. Remember: People are more likely to recall the points you make at these natural places of emphasis.

Choosing an Organizational Pattern

With these basic considerations in mind, you can then review the most common organizational patterns to see which approach seems most appropriate to your purpose, your audience, and the content. Often the content of a particular kind of business writing clearly suggests a particular pattern as most appropriate; at times a mixture of several patterns may be necessary.

A trip report generally calls for a chronological pattern, while an explanation of a mechanical process requires a sequential pattern of a slightly different kind. To choose effectively, you need to review the possibilities. Here are some effective organizational patterns appropriate to readers' expectations and well-suited to presenting certain kinds of information.

DEDUCTIVE ORDER Presenting a general conclusion first, supported by facts and details, allows the reader to get what he or she needs quickly.

INDUCTIVE ORDER Offering specific facts leading to a general conclusion is especially useful when the reader is prejudiced against your position at the start but may be led to see that your conclusion is logical in view of the facts. A common application of this pattern is the problem-solution approach often used in proposals. We will have more to say on this approach in Chapter 8.

ORDER OF IMPORTANCE An ordering of most important to least important facts is useful for business writing because it immediately presents the decision maker with needed information. A least-to-most important factual ordering may be very effective in an oral presentation, however, since it places the most important point at the climax of the presentation, thus leaving a strong impression on the memory of the listeners.

STEP-BY-STEP A sequential pattern is the most effective way of handling instructions or presenting an explanation of a process or procedure.

CHRONOLOGICAL A sequential pattern according to time is useful in many reports, such as trip reports, trouble reports, or schedules.

SPATIAL Presenting information according to physical arrangement is appropriate for a site report, a survey of geographical markets, or certain proposals. The description may involve *direction* (east to west), *dimension* (height, width, length), *shape* (rectangular, circular), or *proportion* (one-half, one-third).

TOPICAL Arranging by topic allows you to cover one topic at a time as a separate and distinct part. It is useful when your topics are independent of one another and not bound in a chronological or logical order. An annual report might serve as an example of this method.

COMPARISON The comparison pattern helps the reader see similarities and differences in the products compared. The products must be compared according to the bases of comparison most relevant to the reader's needs. The whole of one may be discussed completely before the other is considered in a separate paragraph. This whole-by-whole approach allows the reader to see and weigh the advantages and disadvantages of each in relation to his or her needs.

A part-by-part approach may be more useful to the decision maker, however, since it allows the reader to consider each basis separately. For example, the IBM PC could be compared to the Apple Macintosh according to the bases most important to the user: memory, software availability, cost, service, and so forth.

CAUSE AND EFFECT This pattern is often used in projecting what effects may be expected as a result of a proposed action. The effect-to-cause pattern is typified in troubleshooting reports. In writing a cause-effect pattern, be very careful not to draw conclusions your evidence does not support. Evidence must be sufficient, relevant, and representative, and you must demonstrate a strong logical connection between cause and effect.

COMBINATION PATTERNS When you choose one of these methods as a likely approach, remember that the patterns do not work exclusively of one another. You may often encounter situations in which you will effectively combine several of these patterns.

Fitting Pattern to Purpose and Audience

"When do I use each?" To answer that question, you must think about purpose and audience (we say again, at the risk of boring you silly), as well as about content. Writing involves choices. There is both comfort and frustration in the fact that there is no *single* choice that can be considered indisputably "right." The "right" choice is right because it meets the needs of the writer's situation, which of course includes purpose, audience, and content of the message.

Consider some typical writing situations. Let's say you're writing a resume. Your purpose is to inform a potential employer of your qualifications. But is that all? Of course not. You also want to persuade that reader that you have the qualifications he or she is looking for and that you would be an asset to the company. You know also that your reader is very busy, skimming quickly through applications and resumes in search of the qualifications needed.

Would it make sense to organize your resume chronologically, beginning with your grade school education? Or would it be better to figure out what would be most important to the reader and start with that? Remember General Principle 2: When you have a neutral reader, start with the most important point and work toward the less important.

Why do so many resumes ignore this approach in favor of a tedious chronology? Because their writers mindlessly assume there is some rigid form for a resume and therefore never consider the need to think about fitting the strategy to the purpose and audience.

Now let's change the problem. Your boss is thinking about a large investment in data processing. She has listened to presentations for a mainframe computer and for the purchase of several microcomputers instead. She wants to make an intelligent decision based on an unbiased analysis and therefore asks you to give her a report that will help her. Obviously, a comparison would be the most appropriate pattern in this instance.

Likewise, some tasks seem immediately to suggest a particular pattern. A set of instructions will probably be organized step-by-step. But if the most important step is "Turn off the power before you start," that will go first! A trip report may be chronological. A layout for a new laboratory would be spatial. A troubleshooting manual might list the effects and their possible causes.

The point of this discussion is that there is no absolute answer to the question "When do I use each?" But an intelligent analysis of the writing situation will enable you to see that you can narrow your choices and see ways of combining several patterns. The table on page 82 illustrates a way of analyzing the writing situation and choosing an appropriate organizational pattern.

Pattern	Purpose	Audience
Deductive	To convey important information, then details	Neutral, favorable
Inductive	To inform and persuade	Resistant, needs persuading
Order of importance (decreasing)	To emphasize the most important first	Busy, decision-oriented
Order of importance (increasing)	To build credibility and to persuade	Opposed to content
Chronological	To inform or explain	Needs time ordering
Step-by-step	To explain how it works	Needs details in sequence
Spatial	To inform or explain physical arrangement	Needs description or physical features
Topical	To inform about categories not related logically	Needs to see items as separate, distinct
Comparison	Provide information for choices between two or more	Needs comparison in relation to criteria used in choosing
Cause and effect	Explain why it happened or will happen	Needs analysis of possible consequences or reason for problem

Format

The format you choose or design can either help you get your message across or create one more barrier between you and your reader. You can make your format work *for* you or *against* you, so don't take formatting for granted.

In the discussion that follows, we will show you what constitutes a good format, give you some models of standard formats for letters and memos, and remind you of the tools you can use to make your format serve your purpose and your audience.

You will remember that in Chapter 2 we emphasized that all readers have several things in common, and that writers must take these things into account. First, foremost, and always, we said, "ALL READERS ARE BUSY." That undeniable fact elevates the importance of format. Your message may be brilliant, but it ends up as waste paper if the format fails to please the reader's eye sufficiently to get the message read. While it's true that a good package can't substitute for poor contents, appearances count more than most people like to admit. Busy people unconsciously reject or delay attending to communications that look as though they will be tedious to read.

As you go through the mail you receive daily, think about what makes a letter attractive to the reader. At the most basic level of sensory appeal, we all like the feel and appearance of good paper and the look of clean typing, framed by generous margins. We have an instinctive reaction against dot matrix printing from a worn-out ribbon, or a long letter squeezed onto one page without decent margins, or a short letter that finishes on the top quarter of the page, leaving a half-inch margin at the top of the page and four inches at the bottom. "Readers expect writing to have a point, and expect the writer to make that point clear," we reminded you in Chapter 2. They are entitled to expect the conventional and attractive standards of the business world in the letters and memos they receive. You can meet their expectations and spare them surprises if you follow one of the standard formats illustrated in this chapter.

The Basic Parts of a Business Letter

Since variations in format usually involve changing the placement of one or more of the parts, let's take a moment first to review the basic parts of a business letter. The company you eventually work for may use some slight variations on the practices described here, but if you are familiar with these forms, you will find it easy to adapt. Unless your instructor prefers otherwise, choose one of these standard formats, and use it for all assignments involving letters or memos.

Heading

If you are writing on stationery with a letterhead, you can center the date two or three lines under the letterhead, or you can place it flush with either the left or right margin. If your paper does not have a letterhead, type your address flush left or just right of center (depending on the letter format), and place the date immediately under it (without abbreviating the month).

Don't include your name here, just your address. Here is an example, using a flush left position for a block format:

```
1151 Mountain Road
Stowe, VT 05672
August 7, 1985
```

Inside Address

Type the inside address flush with the left margin two to seven lines below the date. Type the individual's name on the first line, the company name on the second, and the rest of the address on the lines that follow.

Salutation

Type the salutation two lines under the inside address. Use the prefix and last name unless you are on a first-name basis with the addressee. Use a colon (:) after the salutation, or else use no punctuation at all. *Never* use a semicolon (;). Use a comma only when you know the addressee well and intend your message as a social letter.

If you don't know the gender of the person addressed, you might want to try an inoffensive salutation like "Dear Customer" or "Dear Parts Manager." Another approach is to use the person's full name. See our discussion of sexist language in Chapter 13 for more help with this occasionally vexing problem of how to address your audience without offending your reader unintentionally.

If you know the gender, the standard prefixes are:

Singular   *Plural*

Mr.	Messrs.
Mrs.	Mmes.
Ms.	Mses. or Mss.
Miss	Misses

Spell out prefixes such as President, Governor, Senator, Professor. Do not use the prefix before the name if it is followed by an abbreviation like M.D., Esq., or Ph.D. We suggest using the simplified letter form, which omits the salutation, if you are writing to a company but do not have the name of a particular recipient. See the example of AMS Simplified Style below. Another possibility, also illustrated below, is the use of an attention line. Either approach is preferable to the use of a greetings such as "Gentlemen" or "Dear Sir or Madam." "To Whom It May Concern" is on its way out as an appropriate salutation; it sounds as though your message were delivered in a bottle that floated in to shore.

Body

Begin the body of the letter two lines below the salutation. Single-space unless the letter is very short; double space between paragraphs. If your letter extends beyond one page, don't use letterhead stationery for the continuation pages. Type a heading six lines below the top of the page consisting of the name of the recipient, the date, and the page number. Then allow four or five spaces before resuming the body of the letter.

Complimentary Close

Placed two lines below the body and aligned with the date, the complimentary close is usually "Yours truly" or "Very truly yours" in formal correspondence. "Sincerely" and "Cordially" are considered less formal. Note that only the first word of the complimentary close is capitalized. Incidentally, if you use "I remain" at the end of the letter to introduce your complimentary close, you are imitating a style of business writing taught a half century ago.

Signature

The signature, which usually appears four lines below the complimentary close, consists of the writer's name, title, and department, if needed. Practice signing your name legibly as well as quickly; contrary to what some seem to think, an illegible signature is not a sign of importance but a sign of rudeness toward the reader.

Additional Parts of a Business Letter

ATTENTION LINE An attention line indicates that you want the person in a particular position (name unknown to you) to read your letter. If you use an attention line, place it two lines below the inside address and two lines above the salutation, as shown below:

```
Penn Body Co.
218 Lake Street
Penn Yan, NY 14527

Attention: Parts Manager

Dear DELCO Customer:
```

TYPIST'S INITIALS If your letter or memo is typed by someone else, it is standard practice to distinguish the typist from the writer in either of the following ways:

```
KJH/jjk
KJH:jjk
```

The first three initials are those of the writer; the second three, those of the typist.

ENCLOSURE NOTATION If you send other documents along with your letter, place an enclosure notation one or two lines below the typist's intials. Use the word "Enclosures" followed by the number of documents enclosed. The abbreviations *Enc* or *Encl* are also used. The purpose of the enclosure notation is to remind the writer what was sent and to allow the recipient to check whether everything was enclosed that was supposed to be.

```
TJC/amk
Encl 2
```

Placement on the Page

By varying the spacing between parts and also by changing your margins, you can lengthen a short letter to improve its appearance on the page or shorten a long letter so you don't end up with one line on the second page. To make a short letter longer, lower the date by as much as five lines, add blank lines between the date and inside address, and allow six blank lines for the signature instead of the usual four.

To shorten a long letter, raise the date to within two lines of the letterhead, and allow only two blank lines between date and inside address.

Plan your margin settings to fit the length of the letter. If your letter contains fewer than 100 words in the body, set your margins at 35 and 75. For an average letter of about 200 words, set them at 30 and 80. If the body of the letter is more than 200 words, set margins at 25 and 85.

Now that you have refreshed your memory on the basics of business formats, let's look at some of the most common formats.

Full Block Format

```
Date

Name of Addressee
Title
Company
Address
City, State ZIP

Dear Ms. Reader:

In Full Block Format, all lines begin flush with the left
margin. There are no indentations for new paragraphs.
Double-space between paragraphs.

This format is preferred by many typists because it
involves no additional margin settings and is therefore
faster. Note that even the complimentary close and the
signature are flush with the left margin.
```

If you are sending copies of the letter to other people,
use a carbon copy notation like the one below. If several
people are receiving copies, list the names alphabetically
or according to level of responsibility.

Sincerely yours,

Writer's Name
Title

WN/fl
cc: Name
        Name

Modified Block Format

                                          Date

Name
Title
Company
Address
City, State ZIP

Dear Customer:

In a Modified Block Format, the date, the complimentary
close, and the signature are aligned slightly to the right
of center on the page. All other lines are flush left.

Some prefer this form to the Full Block because it lends
more balance to the page. It does, however, require more
attention to margin settings.

This model also demonstrates the use of an Enclosure
Notation if materials are to be sent with the letter. The
word "Enclosure(s)" appears directly below the references
initials identifying writer and typist.

                                  Very truly yours,

                                  Writer's Name
                                  Title

JJK/amk
Enclosure: preliminary invoice

AMS (Administrative Management Society) Simplified Format

```
Date

Named Client
Title
Company
Address
City, State ZIP

SUBJECT LINE REPLACES SALUTATION (TYPED IN CAPS)

The AMS Simplified Format streamlines business letters by
eliminating the salutation and the complimentary close. In
so doing, it also eliminates many of the problems we
encounter in searching for nonsexist forms of address for
recipients whose gender is not known to the writer. For
this reason alone it is likely to grow more popular,
although some writers find it too impersonal. Including
the reader's name in the first sentence may be one way of
lessening the impersonality of this format.

As in the Block Format, type all lines flush with the left
margin. Leave six lines between the date and the
letterhead. The inside address appears four lines under
the date, and the subject line, which is typed in
capitals, appears three lines below the inside address and
above the body.

The writer's name and title appear in caps four lines
below the body of the letter.

WRITER'S NAME
TITLE
```

Memo Format

Memos are the principal means of written communication within an organization. Many organizations, therefore, have printed forms containing the necessary headings. If you are using a blank piece of paper, your memo should include the name and title of the recipient, the name and title of the sender, a subject line, and the date. Memos do not require a signature, but the sender may wish to authenticate the memo by initialing it in ink next to his or her name. If the memo is an informal report, it may include several headings to help the readers see the organizational plan.

Memo formats vary slightly from one organization to another. One acceptable format is illustrated below; some writers prefer to place the date flush right at the top of the page. The only really important thing to remember is to be sure to include all four parts of the heading: *to, from, subject,* and *date.*

```
TO:       Personnel Department Employees

FROM:     E. C. Mollenhauer, Supervisor

DATE:     September 1, 1986

SUBJECT:  Use of Consistent Form in Memos


Since the memo constitutes the principal form of written
communication within a company, many companies use a
printed form for such interoffice communications. This
example illustrates the typical format if the memo is
typed on plain paper.

Type all lines flush left. Leave about two inches from the
top of the page before starting your headings. The heading
includes the name of the recipient, the name of the
sender, the date, and the subject of the memo. Do not omit
any of these important items. Some typists prefer to
double-space between each item and to allow three or four
spaces before starting the body of the memo.

You do not need to sign a memo, but some writers and
companies have a practice of initialing next to the
sender's name.
```

Tools for Formatting

Up to this point, we have seen how to make a message visually attractive and clear by using one of the standard formats. But you can do much more if you use a little imagination in formatting your writings. You can use format to emphasize important points, express complicated ideas more simply and economically, and convey logical relationships more effectively.

The tools you have at your disposal for formatting are more varied than you may realize. Once we have the habit of writing mainly in paragraphs, as we do in most academic writing, it just doesn't occur to us to think of using these tools.

Part of learning to be a good business writer is learning to use *all* of the PAFEO tools—and that includes format. We have mentioned many of these tools in our PAFEO process analyses of writing problems throughout this book, but let's look through the tool box once more, concentrating on just those tools that will help us do visually what we intend to do verbally.

Headings

Get in the habit of using headings. They're enormously helpful to readers. For example, readers who are not interested in all of a report can use headings to find just the parts that interest them, and they can later make use of those headings to refer to a particular section. Given the way they are used, you can see that it's extremely important to write headings that are accurate and informative, not cute and misleading.

Headings can even help you as a writer. If you put in the headings after you have completed writing the draft, you will be able to see the overall organization more clearly. Read in sequence, good informative headings and subheadings will form an outline of the whole. Putting in headings after you have finished writing will also help you avoid overuse. Some writers make the mistake of constructing such a complex structure of headings that they destroy the cohesiveness of the whole report. Here, as in all things, moderation and common sense make good guides.

Generally, you will not often need more than three levels for headings. In the absence of specific company format, you may find the following standard format useful in helping distinguish the three level of headings.

1. *Level 1 Head:* All capitals, typed flush left on a line by itself, and separated by one line from the material that follows the head.
2. *Level 2 Head:* Capitals and lower case letters, also set flush left on a separate line and separated by one line from the material that follows.
3. *Level 3 Head:* Capitals and lower case, but typed on the same line with the material it introduces and followed by a period. It may be underlined or italicized to help it stand out from the body of the text.

Lists

Vertical lists change the pace for the reader and are therefore good for emphasizing important points or summarizing points already made.

- Use numbers to rank the items in your list or show sequence.
- Use bullets (·) when ranking or sequencing are not important to your purpose.
- Be sure the items in your list are of comparable importance and length.

Be sure to maintain parallel structure within the list. Use similar word order or construction for similar ideas. If you begin one item with a noun, begin the others the same way. Don't use a sentence fragment for one item and a complete sentence in the next. In the following extract from a letter from the League of Women Voters,[2] note how the writer has listed the organization's position on several public issues, using asterisks for emphasis. Each sentence begins "The League . . ." followed by a verb; that's what we mean by parallel construction. For more on parallel construction, see the discussion in Chapter 14.

** The League strenuously opposed the weakening of affirmative action requirements for federal contractors, as sought by the Office of Federal Contract Compliance Programs, because of the disastrous impact on women and minorities whose employment horizons had just begun to widen.

** The League testified before the Nuclear Regulatory Commission, addressing the lack of comprehensible information on nuclear reactor safety issues, the unresponsive attitudes of NRC staff and the perception that the Commission and regulated utilities work together so closely that the outcome of licensing decisions seems predetermined.

** The League is launching a two-year program to monitor state compliance with the Voting Rights Act, following up on a 1982 pilot project that uncovered resistance to reforms.

Typography

Today's word processors and electronic typewriters allow writers much more flexibility with typefaces than did older typewriters. In addition to permitting underlining, many word processing programs also allow you to switch to italics or boldface type. By using indentations and more white space, you can draw the reader into seeing the most important points, even when he or she is skimming quickly over the page.

In summary, we urge you to think about the possibilities offered by your format, remembering its interrelation with other PAFEO parts. To be effective, the format must serve your purpose and the needs of your reader. It must outline the organizational structures and make the evidence easier to see. How you use formatting tools is limited only by your imagination. When you begin to think of the visual effect of your message as well as its verbal effect, you find yourself adding a new dimension to your communication skills.

Exercises

1. Which of the organizational patterns discussed in this chapter would you prefer in each of the following situations? Why?

[2] Reprinted by permission from the League of Women Voters of the United States.

    a. A routine weekly sales report to your manager.

    b. A progress report on a construction job.

    c. A troubleshooting report.

    d. A procedure explaining how to apply for a grant.

    e. Directions for using a word-processing program.

    f. A memo to your boss summarizing conditions in the home computer industry.

    g. A routine trip report.

    h. The minutes of a meeting.

    i. A corporate annual report.

    j. A report on two possible sites for purchase.

    k. A projection about a future situation after a merger.

2. In 1985, The Coca-Cola Company created quite a stir when it announced it was changing the formula by which Coca-Cola had been made. Then, in response to massive public reaction, The Coca-Cola Company elected to market both "new" Coke and "old" Coke (now called "Coca-Cola classic"), as well as its other soft drink products.

    For purposes of discussion, suppose you have been given the job of writing a memo to the president of The Coca-Cola Company suggesting that the company change the original formula. Then, after the public reaction, suppose you had to write again, now suggesting that The Coca-Cola Company market either (1) both Coke and Coca-Cola classic, or (2) simply Coca-Cola classic. How would you organize each memo? Why? Prepare an outline for each memo using the suggested patterns of organization found in this chapter. ("Coca-Cola" and "Coke" are trademarks of The Coca-Cola Company.)

3. Your company will place a want ad in the local paper to fill an available secretarial position. The evidence for the advertisement is listed below in rather jumbled fashion. Organize and format this evidence to produce a visually attractive advertisement. Assume the advertisement will fill an eighth of a page in the newspaper. Your version, on standard eight-and-a-half-by-eleven-inch paper, will be reduced for the purposes of printing.

    a. Interested applicants should send a cover letter and resume to C. Diment Ronald, Director of Personnel, at your company.

    b. Applicants should have accurate typing skills.

    c. Applicants should have some bookkeeping experience.

    d. Applicants receive a two-week vacation for their first year of work and are paid for all state and federal holidays.

    e. Applicants must be able to type 50 words per minute or better.

    f. The job requires attention to details, initiative, and ability to determine priorities.

    g. Applicants should be outgoing.

    h. Overtime is available.

    i. The salary range is $12,500–$14,000, depending on experience.

    j. The company pays tuition benefits for employees taking college-level courses.

    k. The company provides life, disability, health, and dental insurance.

    l. Applications must be received no later than three weeks from the date the advertisement appears in the newspaper.

    m. The job requires strong organizational abilities.

4. The discount brokerage service of a large commercial bank included the following list on a flyer advertising the advantages of the bank's brokerage service.

   The writer wisely used a list of bulleted items to pace the information for the reader. The writer has problems, however, with parallel structure. Rewrite the list fixing the faulty parallelism. The items can all be made parallel in several different ways.

   - Save up to 70 percent or more on commissions for traded stocks, corporate bonds, and listed options.
   - Prompt execution of your trading orders.
   - You can have the sales proceeds transferred to your checking, savings, Money Market, or NOW account. (You designate one of these accounts as your settlement account.)
   - Securities purchased can be deducted from your settlement account.
   - Stock quotes and up-to-the-minute market data.
   - Cash, option, and margin accounts.
   - Safekeeping of your securities through our clearing broker.
   - Written confirmations of your trade.
   - A toll-free trading number is provided.
   - Detailed monthly statements for active accounts.
   - Access to purchasing government securities, municipal bonds, negotiable certificates of deposit, and more through our Treasury Department.

5. Readers find nothing more deadly than a solid block of prose. To borrow a term from the computer industry, long solid blocks of prose are not "reader friendly." Using the techniques for organizing and formatting that we discussed in this chapter, rewrite the following memo making it more "reader friendly."

```
DATE:      October 15, 1987
TO:        Assistant Managers/Division 87B
FROM:      T. Garnett
SUBJECT:   Cost Overruns/1987 Budget


I have just received the semiannual report that reflects
our expenses for the first six months of this year. While
we are under budget overall, there are three areas where
tighter controls in spending must be exercised. These
three areas are overtime, stationery and supplies, and
insured and uninsured losses. I have attached a memo from
Personnel outlining company policy on overtime
compensation. Please be sure that this policy is being
administered correctly. Ideally, I would like to keep
overtime to a minimum, but I realize that overtime may
prove necessary in light of increasing system problems,
increasing high volumes of work, and end-of-month
processing. It is, however, our job as managers to monitor
the staffing needs we have in order to complete all work
in a timely and accurate manner. If other resources can be
used, do not hesitate to use them. Remember, it is up to
```

you to maintain morale within your units when the need for
overtime arises. Employees should be informed of any
changes in the work routine. If they understand why these
changes are happening and what their involvement is in
implementing change, there will be less resistance and
less chance of decreasing morale. I am looking for you to
reduce the amount of money spent on overtime. Be assured
that I shall work with you to keep overtime hours in our
division to a minimum. Regarding the stationery and supply
expense, I would like you to monitor the supplies on hand
very closely and to order only what is absolutely
necessary. Do not overstock if you can avoid doing so. We
are currently overbudget $3,300.00 in this area. I want to
see this figure reduced dramatically in the coming months.
The category of insured and uninsured losses represents
our biggest variance to date. We are overbudget $11,650.00
in this area. You must do whatever is necessary to ensure
that the bookkeepers follow all procedures for the
disposition of checks. Encourage them to question all
suspicious transactions. Impress upon them the importance
of making the correct decision, and remind them of the
consequences should "bad" checks be paid. We do not have
the luxury of sustaining any further losses in this area
in the coming months. I realize that it is a difficult job
to ensure that no losses result from bookkeeping errors,
but we must do all that we can to prevent them. If
stronger disciplinary action is needed, then we will take
it. If you feel that the bookkeepers would benefit from a
training session to reinforce procedures for handling
these situations, we can easily schedule such a session in
the next few months. I am anxious to hear your suggestions
and comments on this problem, and I look for a positive
turnabout in the next six months so that we will keep our
division underbudget.

6. Here's a letter you'd rather not receive. It's an invitation from the IRS to an
audit, although that nasty word is not mentioned in the letter. As government
communications go, this letter has much to recommend it in its handling of a
difficult subject. Nevertheless, we think a bit more imagination could have helped
greatly in making the format and the organization serve the purpose and the
audience more effectively. Using the tools we've talked about in this chapter,
see if you can make this letter more attractive and easier to read—and thereby
more effective.

Internal Revenue Service
District Director

Department of the Treasury

Date: 11 FEB 1987

Tax Year(s): *1985*

Day and Date of Appointment: *3-4-87*

Time: *11:00 AM*

SS# 111-11-1111
Jane and John Q. Taxpayer
1 Any Street
Anytown, Anystate 00000

Place of Appointment:

Room Number:

5th Floor, Room 5426
600 Arch Street
Philadelphia, Pa. 19106
Exam. Br.—OAS

Contact Telephone Number: Telephone Number 597-1237/86

Appointment Clerk: *OAS*

We are examining your Federal income tax return for the above year(s) and find we need additional information to verify your correct tax. We have, therefore, scheduled the above appointment for you.

If you filed a joint return, either you or your spouse may keep the appointment or you may have an attorney, a certified public accountant, an individual enrolled to practice before the Internal Revenue Service, or a qualified unenrolled individual represent or accompany you. If you are not present, however, your representative must have written authorization to represent you. Form 2848-D, Authorization and Declaration, may be used for this purpose and if your representative does not have copies of this form, they may be obtained from one of our offices. Also, any other individual, even though not qualified to represent you, may accompany you as a witness and assist in establishing the facts in your case.

About the records needed to examine your return—

We would appreciate your bringing to our office the records you used as a basis for the items checked at the end of this letter so we can discuss them with you.

The enclosed Information Guides will help you decide what records to bring. It will save you time if you keep together the records related to each item. Please bring this letter also.

The law requires taxpayers to substantiate all items affecting their tax liabilities when requested to do so. If you do not keep this appointment or do not arrange another, we will have to proceed on the basis of available return information.

About the examination and your appeal rights—

We realize some taxpayers may be concerned about an examination of their tax returns. We hope we can relieve any concern you may have by briefly explaining why we examine, what our procedures are, and what your appeal rights are if you do not agree with the results.

We examine returns to verify the correctness of income, exemptions, credits, and deductions. We find that the vast majority of taxpayers are honest and have nothing to fear from an examination of their tax returns. An examination of such a taxpayer's return does not suggest a suspicion of dishonesty or criminal liability. In many cases, no change is made to the tax liability reported or the taxpayer receives a refund. However, if taxpayers do not substantiate items when requested, we have to act on available return information that may be incomplete. That is why your cooperation is so important.

We will go over your return and records and then explain any proposals to change your tax liability. We want you to understand fully any recommended increase or decrease in your tax, so please don't hesitate to ask questions about anything not clear to you.

If changes are recommended and you agree with them, we will ask you to sign an agreement form. By signing you will indicate your agreement to the amount shown on the form as additional tax you owe, or as a refund due you, and simplify closing your case.

Most people agree with our proposals, and we believe this is because they find our examiners to be fair. But you don't have to agree. If you choose, we can easily arrange to have your case given further consideration. You need only tell the examiner you want to discuss the issue informally with a supervisor, and we will do our best to arrange a meeting immediately.

In addition, you may request the Office of Regional Director of Appeals, which is separate from the district office, to consider your case. We will be glad to explain this procedure and also how to appeal outside the Service to the courts.

We will also be glad to furnish you a copy of our Publication 556, Audit of Returns, Appeal Rights, and Claims for Refund, which explains in detail our procedures covering examinations of tax returns and appeal rights. You can get a copy of this publication by writing us for it or by asking for it when you come to our office.

About repetitive examinations—

We try to avoid unnecessary repetitive examinations of the same items, but this occasionally happens. Therefore, if your tax return was examined in either of the 2 previous years for the same items checked on this letter and the examination resulted in no change to your tax liability, please notify the appointment clerk as soon as possible. The examination of your return will then be suspended pending a review of our files to determine whether it should proceed.

About your appointment—

Your appointment is the next step unless, of course, you notify us of a repetitive examination as outlined in the preceding paragraph. If the date or time of the appointment is inconvenient, please call the appointment clerk to arrange a more suitable time. We will consider the appointment confirmed if we do not hear from you at least 7 days before the scheduled date.

If you have any questions, please contact the appointment clerk whose name and telephone number are shown in the heading of this letter.

Thank you for your cooperation.

Sincerely yours,

District Director

Enclosures:
Information Guides
Publication 876, Privacy Act Notice

CHAPTER 5

Revision

Our combined teaching experience totals almost half a century. During that time, we have discussed the writing process with thousands of our students. Much of what we have written in this book has grown out of those discussions. We suspect that students who have serious writing problems have even more serious problems at the other two stages: they lack strategies for planning, and they don't know how to revise systematically. For the first problem, we offer the PAFEO remedy. For the second problem, we offer some important distinctions and some useful tips in this chapter.

How *Not* to Revise

Here is a scene we have both replayed hundreds of times in our offices. It may sound familiar to you too. We present it to you as an illustration of the problem commonly confronted by those who hear the urgent advice, "Revise!" but only find that advice confusing and upsetting. Perhaps you will understand how this student feels.

"Gerry, tell me all about the process you followed in writing this paper."

"Like what do you mean? I didn't copy it or anything. I wrote a rough draft, and then I revised it just like you said we should."

"Tell me about how you did that revising. That's what I'm interested in now."

"Well, first I got my roommate to read it over. His sister's an English major. He didn't like some stuff, so I changed it. Then I read it over. I mean like real carefully. Word for word. And I changed some more stuff and put in the commas where I thought they belonged. And I had to look up a couple of words I wasn't sure how to spell. And then I typed up my good copy. I changed a few words around when I was typing it. They just didn't sound right."

If your revising process is anything like Gerry's, you ought to be able to pick up some useful techniques in this chapter. Note, first of all, some of the assumptions behind what Gerry is saying. He assumes that his first draft is pretty close to being his last draft. It needs revision, of course, but Gerry assumes revision means tinkering with the wording and correcting the spelling. He assumes that revision is something you can do in one pass through the paper, providing you read carefully enough. He also seems to assume that changes in what doesn't "sound right" will be for the better, even if he hasn't tried to diagnose *why* it doesn't sound right.

Gerry's predicament is widely shared by inexperienced writers. Everybody says revision is good for what ails your writing, but nobody tells you how you are supposed to go about it. Professionals make their living on their revising skills. They revise their own work and other people's work, and they don't even have roommates whose sisters are English majors. They have something better: an understanding that writing is written to be revised, and a systematic approach that separates revising techniques from editing techniques.

Now that you know what doesn't work and why, let's examine the way professionals practice the all-important art of revising. To begin with, their attitude toward revising differs from that of most students. Many students see revision as acknowledging failure. They have failed to do the thing right the first time, so now they must do it again. Experienced writers, on the other hand, expect to revise. They don't see it as a separate process to be saved for the final draft. They see it as part of the ongoing process of writing. They're always rethinking ideas, tearing things apart, moving parts around, trying different approaches.

If you can learn to see revision as an opportunity to get closer to your purpose and your audience, not as a fix-up chore, you too will be writing in a more professional way.

Once you have accepted the idea that revision is a necessary part of writing, you will have to plan accordingly. Make it easy for yourself. If you write, skip lines. If you type, triple space. If you use a word processor, give thanks for the built-in ease of revising. If you write, be sure to write on only one side of the page. You can easily cut and paste paragraphs into new arrangements if there is nothing on the reverse side. Develop your own shorthand for telling yourself what you need to do. Use carets (^) where you want to insert words, for example, and single lines to cross out words or passages to be omitted. (Don't black them out so you can't read them; it's a writer's privilege to change his or her mind.) Use arrows to indicate relocations, or renumber paragraphs in the margin if you like. Revising is a personal matter. Find the system that suits you best.

Using the PAFEO Checklist

When you first read through a draft, concentrate on the ideas, the organization, the connections that show relationships. Is your meaning coming through? Could you outline your structure if you had to? When you encounter a trouble spot, ask yourself, "Just what am I trying to say here?" and then ask, "What's stopping me?"

You can use the PAFEO Process to help you focus on the big picture before you get caught up in editing. Revising ought to go deeper than surface corrections if it is to do any good. First read your draft for ideas, organization, and relationships. See if your overall meaning and plan emerges clearly. The following checklist can help you see the forest before you wander among the trees.

The PAFEO Checklist

Purpose

- Is the purpose of my memo, letter, or report clear to the reader?
- Does it meet the purpose intended by the person who requested it?
- Do the things that I want the reader to remember stand out clearly?

Audience

- Who is going to be my principal reader, the one who will act on this? Will it also be read by others? Does what I have written meet the needs of these separate audiences, or must I address those needs in two separate communications?
- Have I provided sufficient background so that reader(s) can understand my message? Have I given them details they don't need and are not interested in? Have I said anything or used technical jargon that will alienate one group of readers?
- Have I identified a goal the reader and I can share? Is my writing reader-based in that it draws conclusions and guides the reader through my organizational plan?
- Have I taken into account my reader's attitude toward me and toward my ideas, and organized accordingly?

Format

- Have I met the reader's expectations in the standard letter, memo, or report forms? Have I remembered the date and used the subject line to give specific information?
- Have I used cues such as headings, white space, typography, and indentation to make it easy for the reader to follow? Do I need to put in or take out some cues because they distract the reader or make for choppy reading?
- Have I supplied appropriate graphics whenever they will be helpful? Have I cited each graphic in the text, labeling each for ease of reference?

Evidence

- Have I checked each fact and each number for accuracy?
- Does each main point have adequate support, and is the evidence relevant and representative?
- Do I have enough to make the reader see what I mean, or do I have so much that I'm belaboring the obvious?

Organization

- Have I organized with the reader's response in mind?
- Is my organizational pattern appropriate to my purpose, my audience, and my message?
- Does what I have written have a beginning that states my purpose, a middle that supports my purpose with specific evidence, and an ending that summarizes or recommends?
- Do I emphasize the most important ideas by putting them at the beginning or the end (or both)?
- Do I have enough transitional words to enable the reader to follow my thoughts?

Focusing on Paragraphs

You will help yourself with your organizational problems if you use each paragraph to group and organize ideas. Purposeful paragraphs help you think in manageable conceptual blocks. Each well-formed paragraph is a miniature version of the whole piece of writing. It has a main idea expressed in a topic sentence. It has supporting evidence organized in a way most appropriate to the needs of the audience. It is unified around a single idea, and the sentences in it relate to each other in a logical or chronological order. Because paragraphs are units of thought, each should relate to the central idea you are writing about.

By focusing on your paragraphs as you edit, you ought to be able to spot gaps in your own thought pattern, gaps that will present obstacles to your reader. As you examine each paragraph, test it for unity and conherence by asking these questions:

1. What does the paragraph do in terms of the whole report (or memo or letter)? If it doesn't do anything, it may be a candidate for cutting. Many a paragraph exists only because the writer had some information and didn't know what to do with it.
2. Does the paragraph follow logically or chronologically from the one preceding it? Does it lead naturally into the one following it?

Focusing on Sentences

Much of what is bad in the style of business writers can be found in their sentence structure. You can usually spot bad sentences: they're swollen like an infected thumb. Here's a paragraph from a letter sent out by a civic group bent on honoring the mayor.

```
We are in preparation of a Souvenir Program Book
acknowledging this fine man and his many achievements
along with a proposed brief history and photographs of our
town, featured advertising from the business community,
together with congratulatory comments and sentiments
expressed by you and the many fraternal organizations that
would like to be included in this "Keepsake Program." We
have taken the liberty, and for your convenience, to
enclose an order form describing this program and
hopefully we ascertain your full cooperation.
```

Do you have a sense that all is not well with this kettle of lard? How in the name of all that's clear and concise can we possbily revise it?

It may not be as hard to revise as it looks. Researchers and teachers of writing agree on some basic but highly effective revision principles.[1] Using these principles can help you change a weak sentence into a healthy one with miraculous ease.

Checklist for Revising Sentences

1. Look for the important action in the sentence and express it in a simple verb if possible. State *who is doing the action* in the subject and *what the subject is doing* in the verb. Or as a musician friend puts it, tell "What's happening?" and "Who's shaking?"
2. Divide overloaded sentences. Average sentence length today is fifteen words (compared to twenty-nine a century ago).

[1] This chapter draws upon observations and advice from many writers in the field. We acknowledge indebtedness particularly to Geraldine Henze, *From Murk to Masterpiece* (Homewood, IL: Irwin, 1984); John Keenan, *Feel Free to Write* (New York: Wiley, 1982); Richard A. Lanham, *Revising Business Prose* (New York: Scribner's, 1981); and Joseph M. Williams, *Style,* 2nd ed. (Glenview, IL: Scott, Foresman, 1985).

3. Change passive voice to active, except when the emphasis is on the receiver, not the doer.
4. Count prepositions (*about, around, at, before, by, down, from, in, near, of, on, over, through, to, toward, under, up, with*). If the sentence has four or more, it resembles a list. Lists have their places, but listlike sentences are often weak because they don't distinguish main ideas from subordinate ideas. To fix the sentence, put subordinate ideas in subordinate clauses and get rid of unnecessary prepositional phrases. Applying the first principle should help eliminate many prepositional phrases.
5. Check each use of the verb *to be* (*is, am, was, were, being, been*) to see whether you can substitute an action verb.
6. Look for strings of noun modifiers and break them up, using Principle 1.
7. Look for abstract and general words and see if you can make them more concrete and specific.
8. Get rid of monotonous sentences by varying the length and structure.
9. Be sure the antecedent of each pronoun is clear and unmistakable.
10. Read the sentence aloud, remembering essayist William Hazlitt's comment: "No style is good that is not fit to be spoken or read aloud with effect."

Now let's see if the checklist can help us revise the civic committee's prose into something closer to their meaning.

The first sentence weighs in at fifty-eight words, including seven prepositional phrases. The important action? They're going to publish a souvenir program. Since the sentence is overloaded, we will divide it and get rid of the passive. Then we'll add a bit of variety to the structure and length of the sentences and try to make the words more concrete. You can test the result by reading the original and the following revision aloud.

> The Dedication Committee intends to publish a souvenir program honoring the mayor's achievements. The program will feature a brief history and some rare old photographs of the town . . . and you. That's right. Picture yourself or your business in this memorable book too. Just enclose your advertisement or congratulatory message with the enclosed order form and your check.

You might revise it differently, and we will offer you the opportunity to do so. But this revision shows how much improvement you can make mainly by shortening sentences, identifying an action and the agent of that action, and making wording more concrete. This revision also attempts to be more reader-based than writer-based. Can you tell how?

Focusing on Words

Since Chapter 13 tackles problems of wording in some detail, we'll confine ourselves here to a checklist you may find helpful in identifying your particular revising needs.

1. Have I used words most likely to be understood by my readers, eliminating inappropriate jargon?
2. Have I used words that convey my exact meaning, checking the dictionary when uncertain?
3. Have I cut irrelevant, repeated, and redundant words?
4. Is my tone appropriate to my audience?
5. Have I eliminated sexist language?

Focusing on Correctness

The final matters you should check are those dealing with correctness (grammar, punctuation, and spelling). We hope you will find Chapter 14 a helpful tool in mastering these areas. Just a word concerning our placement of these matters: Because we mention them last is no indication that we consider them unimportant. *All* revisions matter greatly, and neglect of one may negate all you've accomplished in other respects.

The one thing you forgot to do is likely to be the first thing your reader notices. Discouraging as that may seem, it should only encourage you to make the most of your revision. Use the checklists in this chapter until you are so familiar with them that you don't need them anymore. By that time, you'll be a good reviser, and a good reviser is a good writer.

Exercises

1. It is important for you to have a true picture of your own revising and editing practices. Think about the last writing assignment you did for this course or one of your other courses. Now write a step-by-step description of the revising process you used. You need not show it to anyone else, unless your instructor asks to see it for diagnostic purposes, but you should read it over to see if you can adopt some of our suggestions to improve your own revising practices.

2. Apply the PAFEO checklist to any writing assignment specified by your instructor. What revisions would you make, if any, as a result of applying the checklist?

3. Revise the following paragraphs from a report prepared by an academic computer center. They could be combined into one paragraph dealing with "Basic Support Services." Use the checklists in this chapter to help you.

BASIC SUPPORT SERVICES

```
Support services are activities designed to provide a
stable computing environment and assist academic users
with computing activities.
```

Currently there are few computer center support services
primarily because the first La Salle academic computing
consultant has just been hired.

Technical support for problem resolution, software
implementation, and software maintenance has been provided
on a "hit-and-miss" basis if at all.

The type of basic support services that academic users
should come to expect are documentation and training for
the DEC and Prime as well as for micros. Consulting for
supported languages, packages, and operating systems also
comprise user services. Technical support includes
academic liaisons; software upgrades, installation, and
testing; and dissemination of information to the user
community.

Basic support services are designed to help people use
academic computing resources more efficiently and
comfortably by providing guidance and orientation to
computing. As technology grows and software is enhanced or
is replaced, computer services should keep the user
community informed.

4. Now it's time for you to try your hand at revising the Dedication Committee's
handiwork. Since your instructor may want you to revise the entire letter, here
it is in its picturesque glory, with names changed to protect the guilty.

Dear Friends and Neighbors:

We are delighted to announce that on Saturday, June 21,
the citizens of our lovely town and other groups will pay
homage to the Honorable Mayor John J. King. This planned
occasion will be the formal dedication of the existing
community center that will subsequently be known as:

                    John J. King Community Center

It is our every wish the dedication ceremonies and
celebration will be the highlights of the annual Straw Hat
Weekend scheduled for that date.

Needless to say, our mayor has distinguished himself as a
public servant to his constituents, neighbors, and
friends, not only as a member of the Board of
Commissioners but as a concerned private citizen for the
past 20 years displaying his extraordinary talents, always
enlisting his vivid imagination when called upon. Hence,
religious, civic, government, and public leaders from many

areas will be joining us on June 21 to honor our
outstanding mayor commensurating this eventful day.

We are in preparation of a Souvenir Program Book
acknowledging this fine man and his many achievements
along with a proposed brief history and photographs of our
town, featured advertising from the business community,
together with congratulatory comments and sentiments
expressed by you and the many fraternal organizations that
would like to be included in this "Keepsake Program." We
have taken the liberty, and for your convenience, to
enclose an order form describing this program and
hopefully we ascertain your full cooperation.

We would be remiss not to mention the undersigned would
graciously welcome you in person on this gala occasion and
are looking forward indeed to that meeting!

Cordially, together with our acknowledging gratitude to
the many affiliations,

N. Morton
M. Rosemont
N. Morton and M. Rosemont
Co-Chairs

5. Once you become familiar with your set of revising tools, you can apply them
   to the reading you do in your own field. For example, pick a likely page from
   one of your own textbooks and revise it. Make a copy of the original, and turn
   it in with your revised version. Be prepared to defend your changes with a more
   specific reason than "It sounds better."
6. Try the same technique with an article from one of the journals in your field.
   You may be surprised and encouraged to see that all that is printed is not perfect.

PART II

PRODUCTS

CHAPTER 6

Resumes and Cover Letters

In Part I of this book, we introduced you to our approach to the writing process, the PAFEO process. We then examined each of the five parts of that process in some detail.

In Part II, we will show you how to apply the PAFEO process to the various products the world of work will require you to write. Our approach in this part of the book is not encyclopedic. We cannot possibly tell you how to write every kind of document you may be called upon to write on the job. Nevertheless, we have tried to be comprehensive. Where they are beneficial, we have provided checklists and models, not as certified directives from on high, but as aids to illustrate possible solutions to writing problems when you don't know where else to turn. In other cases, we have left our discussions more general, because not all business writing situations lend themselves to the kind of quick fix a checklist can sometimes provide. In all cases, we have tried to provide problem-solving strategies that, when combined with your own common sense, should see you through as many difficult writing situations as possible.

PAFEO and the Job Search: An Overview

We begin our discussion of product with a chapter on the writing of resumes, cover letters, and other documents concerned with securing employment.

Our first advice is to read what follows carefully, read everything else you can get your hands on carefully, talk to as many people as possible, and then disregard whatever will not work for you.

There is no one correct way to put together a resume or a cover letter. But there are resumes and cover letters that do excellent jobs of selling what applicants have to offer, and there are resumes and cover letters that tell busy recruiting managers absolutely *nothing* that they need to know.

One of our former students, now a recruiting manager, recently received the following unsolicited resume along with a letter asking to be "considered for any available openings." While we have changed some particulars so as not to embarass the applicant, it is apparent that she will have a difficult time attracting much interest in her application with a resume that says as little as hers does.

```
Wilma Bryant
6754 Lintel Street
Bethlehem, PA 18019
876-5555

Education          Graduate of Bethlehem Regional
                   High School

Qualifications     30 years in administration for
                   100 people in various
                   departments

Reference          Horne-Naples Manufacturing Co.,
                   Inc.
                   P.O. Box 345
                   Bethlehem, PA 18034
```

Such a resume gives a recruiting manager little difficulty; it makes its way rather quickly into the "circular file."

In the pages that follow, we will briefly discuss how you can apply the five steps in the PAFEO process to writing resumes and cover letters. Then we will consider the issue of evidence. In talking with students over the years, we have found two recurring problems: (1) students think that they have nothing of interest to a potential employer to put in a resume or a cover letter, or (2) having figured out what they want to tell potential employers, students don't know how to present that information.

After discussing the issue of evidence in some detail, we will provide some sample resumes and cover letters and then comment briefly on additional situations related to getting a job.

Purpose

Your purpose in writing a resume and cover letter is simple: you want an interview for a job. No one lands a position solely on the basis of a resume and cover letter—unless you are going to work for the family-owned firm.

In your resume and cover letter, you want to convince a potential employer that it would be worthwhile to sit down with you face to face to discuss what you have to offer.

Experienced recruiting managers know exactly what to look for in a resume or cover letter. You shouldn't try to bluff them. You can't run the risk of confusing them. You can't waste their time. And they will give you very little time. In a fast ten seconds, recruiting managers will have read everything they need to read to reject your resume and cover letter outright or to pursue them further.

Audience

The resume and cover letter you send will face one of the most difficult audiences imaginable: an experienced recruiting manager. In his excellent guide for college students seeking jobs, *Jobs for English Majors and Other Smart People,* John L. Munschauer, Director of Career Development Services at Cornell University, relates the following story:

> To find out about the effectiveness of letters and resumes, I visited corporations and asked employment managers for their comments. "Here," said one employment manager, as he picked up an eighteen-inch stack of letters and handed it to me. "This is my morning's mail. Read these letters and you'll have your answer."
>
> "I can't," I protested, "I have only two hours, and there's a day's reading here."
>
> "Yes, you can," he replied. "Unfortunately, you'll get through the pack in a half hour because a glance will tell you that most are not worth reading."
>
> It was hard to believe that letters could be that bad, but he was right. The typical letter was an insult.[1]

In talking with recruiting managers, we discovered that companies may follow two different general procedures in screening resumes and cover

[1] John L. Munschauer, *Jobs for English Majors and Other Smart People* (Princeton: Peterson's Guides, 1982) pp. 95–96.

letters. In some companies, all initial screening is done by someone specifically trained in personnel work; in others, initial screening is done by the actual manager for whom the candidate will work. In one case, then, you are directing your resume and cover letter to a generalist. In the other, you are directing them to a specialist in the field in which you wish to work. In a third case, you could be writing both to the generalist with a background in personnel and to a specialist with a background in some specific technical area. Whether writing to a generalist or a specialist, you need to tell your reader what he or she needs and wants to know.

Our own university's Career Planning and Placement Bureau recently sponsored a panel featuring the recruiting managers from four companies that had hired the most La Salle University graduates in the past few years. All four managers stressed that they looked at resumes and cover letters to determine as quickly as possible what preparation or experience candidates had in the following skills and areas:

- Communications skills (both written and oral).
- Interpersonal relations as demonstrated by the ability to work as a member of a team.
- An entrepreneurial awareness, a sense of what the world of work involves and demands.
- Specific skills in one technical area and secondary skills in a variety of related technical areas.
- Basic computer literacy (for jobs not specifically in the area of information science).
- A sense of business ethics.
- The ability to manage time, set priorities, and work under stress.

While no one candidate can be expected to excel in all these skills and areas, this list does provide some guidelines for candidates communicating with recruiting managers by letter or resume or in person during an interview.

Format

With such highly critical audiences to face, you would think that job applicants would give some care to what they have to say about themselves and how they say it. Experience often proves the contrary. Job applicants, out of ignorance more often than not, commit "suicide by letter bomb" in their cover letters and resumes.[2]

Sloppy, ill-prepared, and poorly proofread application materials pour in daily to recruiting managers. Some excerpts from these materials appear in the exercises at the end of this chapter.

[2] We came across this characterization of job application materials in, of all places, an article lamenting the quality of letters sent by applicants, many holding Ph.D.s, for teaching positions in English. See Lawrence H. Martin, Jr., "Suicide by Letter Bomb," *ADE Bulletin*, 67 (Spring 1981), 22–25.

Your resume and cover letter must be typed; in some cases, you may want to have them typeset. Some ill-informed or lazy applicants handwrite their materials in pen, even in pencil. Others misspell the recruiting manager's name, innocently ask for international jobs with companies that have only local offices, or make outrageous claims about their past experiences and training. Don't make these mistakes. Be effective and professional in your presentation and your content.

Generally, your cover letter will fill one page, as will your resume. A resume only a half page long may suggest you are only half a person. When the quantity of your evidence dictates, your resume can run to a second page.

You must be prepared to do your homework and to spend a great deal of time carefully organizing, editing, and proofreading your resume and cover letter. You are being given the opportunity to "sell" yourself through these materials—make the most of this opportunity.

Evidence

As we indicated earlier, the greater part of this chapter will discuss the issue of evidence as it applies to resumes and cover letters. Some initial comments are in order.

First, don't lie. Liars are never very imaginative; invariably they indicate they hold advanced degrees from big name schools—a fact easily verified by a potential employer. Or they indicate they held major responsibilities with a major company or a competitor—another fact easily verified.

Second, don't say negative things about yourself. At times, our own students have an amazing penchant for saying negative things about themselves. When responding to job advertisements asking for candidates with courses in computer science, for instance, they unhesitatingly admit to having had no computer courses. By so doing, they may have closed off a potential area for discussion in an interview or had their application rejected immediately. Many of our students though, without ever having had a college-level course in computer science, have worked with computers or held jobs that required basic word-processing skills. We don't counsel that these students lie and say that they have had courses in computer science. We do counsel that they turn such issues to their advantage by looking at themselves from a positive rather than a negative point of view. Recruiting managers basically want to know what you have to offer. Tell them without apology and by accentuating the positive in your evidence.

Above all, in presenting evidence, use common sense. Certain types of evidence commonly listed or requested in job applications in the past are now illegal for employers to ask: your age, race, sex, marital status, number of children you have, and details about any legal difficulties you may have had. For most positions, health status is also no longer requested.

If you are in college, or if you have just graduated from college, what you did in high school and grade school is not generally going to be of much interest to a recruiting manager. A recruiting manager does want to know where you went to college, what you majored in, possibly what courses you took, and what honors or distinctions you may have earned. The recruiting manager wants to know where you worked, how long you worked there, and what you did on your previous job. The recruiting manager wants to know anything else about you that might be an asset for a potential employee. Awards, hobbies, and skills—as long as they have an application to the job you are seeking—are examples of such assets.

Most importantly, the recruiting manager needs to know your name, mailing address, and telephone number. The last two are mysteriously absent from a surprising number of resumes and cover letters.

By the way, we know a recruiting manager who advises job candidates to purchase a telephone answering machine. He argues that not only does he need a job candidate's telephone number but that he also needs to have his telephone call answered so he can leave messages for job candidates he might be interested in interviewing.

Organization

Traditional, or chronological, resumes begin with your name, mailing address, and telephone number. Follow such initial information with data on your education, previous work experience, and other items of relevance to the job for which you are applying. In presenting these data, you should use whatever organizational pattern best suits your credentials. Some resumes lead off with education; others, with previous work experience.

Functional resumes also begin with your name, address, and telephone number, but then present remaining data according to the skills you have to offer.

In both types of resumes, you can include an objective related to your immediate job search or your long-range career plans. For the sake of clarity, arrange data consistently with an eye to parallel structure. Separate points clearly; present educational and work data in reverse order, beginning with the most recent experience and working backwards. (See the sample resumes on pages 128–135.)

Evidence: Making Sure You Have It When You Need It

You cannot wait until the week before graduation to begin your job search. The week before graduation is no time to discover that you don't have much to offer a potential employer.

At the risk of being accused of reinforcing the overemphasis on careerism in American higher education today, we would like to suggest that you begin preparing for your job search as early in your college career as possible.

Unless you are a graduating senior, you can better position yourself for a job search while still in school in two ways: course selection and extracurricular activities. Our own students in business writing courses are just as likely to be sophomores and juniors as they are to be graduating seniors. Their majors vary from the humanities, to the sciences, to the fields of business.

To students majoring in the humanities, we suggest balancing courses that teach them to read, to write, and to think with courses in business, science, and information systems, where they can combine their reading, writing, and thinking skills with some background basic to the world of work.

To students majoring in the sciences or business, we suggest balancing courses that teach technical skills with courses in the humanities. Even recruiting managers are becoming increasingly hostile to accountants who can only tally balance sheets and scientists who can only scribble equations. What the world of business and industry wants are managers, technicians, and support personnel with solid foundations in their majors balanced by diversification and flexibility in elective and supportive courses. The ability to communicate effectively has long been high on the list of qualities that characterize promotable executives.[3]

No matter what your major, look into the possibilities of work-study jobs, internships, co-op positions, and even volunteer activities that will provide you with career-related evidence for your resume and cover letter.

Extracurricular activities can also help you better position yourself in the job market. You don't necessarily need to be captain of the basketball team, sorority president, or student senator, but do take part in some extracurricular activities. Depending on the activity, your participation can show a recruiting manager that you are more than a bookworm and that you are a potentially good team player. Every college and university sponsors a variety of activities and organizations. Find out what your school has to offer and become involved.

Most professions have membership associations that admit students or have student auxiliaries. Find out about these groups, join them, and take part in their activities. Such initiative on your part may provide you with valuable tips and contacts for jobs, as well as an entry for your resume that might just set it apart from the other 300 in the stack.

You can prepare early for a career in a third way by talking to people before your senior year. First on your list of people to talk to should be the staffs of your school's counseling service and placement office. These

---

[3] Garda W. Bowman, "What Helps or Harms Promotability?" *Harvard Business Review,* 42 (January–February, 1964), 14.

people are paid to help you; use them. Second, talk to your professors, especially professors who have active contacts off campus in the business community. Many professors do part-time consulting, which gives them access to the companies and agencies that might be interested in hiring you. Find out from your parents and their friends who does the hiring where they work, and then talk to these people.

You may also be able to talk to people in businesses and industries where you would like to work without using your professors, parents, or friends as go-betweens. Some people are very generous with their time and are glad to help beginners. If you are able to set up a formal or informal meeting or interview, be prepared to ask specific questions and to listen to sound advice.

Finally, read newspapers, journals, and magazines in the fields of employment that interest you so that you can stay abreast of developments. Also, check out career opportunities at your school's placement office, counseling center, or library.

In short, to prepare for a job after graduation, start early, be energetic, be resourceful, and be persistent. You can do all these things and still enjoy your college years.

Evidence: Making Sure You Present It To Your Audience Effectively

In his excellent guide to the job search, *Jobs for English Majors and Other Smart People,* which we mentioned earlier in this chapter, John Munschauer advises students and job seekers "to listen to the language of employers."[4] We would like to suggest a warm-up exercise you can carry out before you sit down to write a resume or cover letter. This exercise is designed to give you some practice in talking the language of employers.

Resumes and cover letters are really nothing more than advertisements for yourself. You must, therefore, know what it is you are trying to sell and how best to sell it to your customer, the recruiting manager. Before you can make your sales pitch, you need to compile an inventory of what you have to offer a prospective employer. The data in this inventory are nothing more than the evidence—the "E" in PAFEO—that you will want on hand when composing your resume and cover letter. This evidence includes:

- Education and training.
- Current and previous work experience.
- Special skills you may have mastered.

---

[4] John L. Munschauer, *Jobs for English Majors and Other Smart People* (Princeton: Peterson's Guides, 1982), pp. 21 ff.

- Participation in professionally related activities.
- Awards or distinctions you may have earned.

Using a Job Description to Focus on Evidence

You can begin compiling data for an inventory of the evidence that you will need by writing a job description for a part- or full-time, volunteer or paid, position you now hold or recently left. (If you are a full-time student and are not holding down a job, remember that studying is a full-time job for which you can also put together a job description.)

Traditionally, job descriptions have been management tools.[5] Managers use them to fill or to redefine positions within their companies. As a job candidate, you can use a job description to provide you with a clear, concrete list of responsibilities and duties.

Formats for job descriptions vary. For the purposes of this exercise, though, include the following information:

- Specific job title.
- Position of the job within the company.
- General responsibilities.
- Specific duties.
- Necessary background for the position.

SPECIFIC JOB TITLE At college and universities, everyone seems to be an assistant dean of some sort. At banks and insurance companies, everyone seems to be an assistant vice-president. Be prepared to identify your present position—and any earlier positions—in a way that will make sense to a recruiting manager not associated with the company for which you presently work. If you are a "Salesperson," what products or services do you sell? Do you work full-time or on a contingency basis? Are you salaried or do you earn commission? If you are "Assistant to the Department Head," how big is the department? Is your position as "Clerk/Typist III" more or less important than that of "Clerk/Typist II" or "Clerk/Typist I"? In some sports, the low score wins, and you cannot expect the recruiting manager from Company X to be familiar with the organizational structure of Company Y.

In a resume or cover letter, you will list job titles. If these titles are not self-explanatory, be prepared beforehand to make them so by adding a brief explanatory or qualifying phrase. Here are some comparative examples of nonspecific and specific job titles. The general area of employment is indicated in italics.

Oil company: Staff Writer vs. Staff Writer, Compensation and Benefits.
Major building firm: Assistant to the Director of Public Relations vs. Assistant to the Director of Public Relations, Northeastern and Canadian Offices.

---

[5] The standard reference on job descriptions from a management point of view remains Joseph Famularo's *Organization Planning Manual* (New York: American Management Association, 1971).

Government office: Service Representative vs. Service Representative, New Claims and Information Services.
College: Assistant Manager vs. Assistant Manager, Reporting and Analysis, Financial Division.
Performing arts complex: Assistant vs. Production Lighting Assistant.
Law firm: Legal Aide vs. Legal Aide in a 24-lawyer firm.

POSITION OF THE JOB WITHIN THE COMPANY Before putting together a resume or cover letter, be prepared to include specifics about whom you report to, both on a day-to-day basis and intermittently.

Here are the same job titles listed above made even more specific by expanding them to include information about accountability within an organizational structure:

Staff Writer reporting directly to the Manager of Compensation and Benefits.
Assistant Director of Public Relations, Northeastern and Canadian Offices, reporting directly to the Chairman and Vice-President for Marketing.
Service Representative, New claims and Information Services, reporting directly to the Office Operations Supervisor.
Assistant Manager, Reporting and Analysis, Financial Division, reporting directly to the Controller and Vice-President for Finance.
Production Lighting Assistant reporting directly to the Technical Director.
Legal Aide in a 24-lawyer firm reporting directly to the three senior partners.

GENERAL RESPONSIBILITIES Few jobs are one-dimensional. Instead, most jobs require people to do a variety of things of varying degrees of importance. In making an inventory of your job skills, list your major responsibilities in descending order of importance or frequency. You may in the future want to tinker with the order in selling yourself for different jobs, but at this point simply make sure you have a specific list of what it is you do as part of your job. Use this list to explain further what your specific job title means.

SPECIFIC DUTIES List your duties so that they will make sense to a recruiting manager. Move from most important or most frequent to least important or infrequent. To get into the habit of selling yourself, begin each duty in your list with an active verb.

Here is a sample statement of a specific duty for each of the expanded job titles just listed:

Ensure that all departmental communications meet federal and state regulatory guidelines.
Edit all promotional guidelines.
Interview applicants to determine their eligibility for benefits.
Manage budget preparation for the university.
Assist in the operation and maintenance of the theater's lighting equipment.
File documents at City Hall or in District Court.

NECESSARY BACKGROUND FOR THE POSITION   Part of your resume and cover letter will list your training and educational background. Be prepared to specify all formal training that prepared you for your present position. Include schooling, in-house training, and any other appropriate learning experiences on and off the job.

Job descriptions can be an effective tool for getting down on paper an outline of what you have to offer a recruiting manager. Put one together as a way of better preparing you to write a resume or cover letter. Here are two samples you may want to follow.

Job Description (Sample 1)

Specific Job Title and Position within the Organizational Structure:	Part-time Legal Aid in a 24-lawyer firm reporting directly to the three senior partners.
General Responsibilities:	Assist the attorneys in their research and maintain the general office and firm library.
Specific Duties:	Oversee the office library. Conduct research at nearby law libraries. File documents at City Hall or in District Court. Maintain office equipment and inventory office supplies. Run occasional errands for staff and clients.
Prerequisites, Training, or Background:	High school diploma. Two years' experience in the field of law.

Job Description (Sample 2)

Specific Job Title and Position within the Organizational Structure:	Assistant Manager, Reporting and Analysis, Financial Division, reporting directly to the Controller and Vice-President.
General Responsibilities:	Assist the Controller in monitoring the statistical and financial activity in the University and in reporting that activity to senior administrators regularly.
Specific Duties:	Prepare all financial and statistical questionnaires. Assist in preparing the University's budget and annual financial report. Assist in monitoring endowment activities. Represent the Controller when necessary at divisional meetings.
Prerequisites, Training, or Background:	Bachelor's degree in Accounting. Minimum of one year in financial management. Excellent written and oral communications skills.

Resumes: Evidence

Let us begin this detailed discussion of what evidence to include on your resume by reminding you that what follows is only a series of suggestions. There is no one correct way of presenting evidence on a resume, but there are resumes that advance an applicant's candidacy, and there are resumes that help applicants commit "suicide by letter bomb."

Your Name, Address, and Telephone Number

First things first. Begin with your name—eventually you will want to make sure it appears in all caps or in bold face print on your final product. You waste time and stress the obvious by beginning with "resume of." If a recruiting manager cannot tell at a glance that this piece of paper is your resume, you have already lost the job. (By the way, leave the accents off the word *resume*—even if you are applying for a job in Quebec or France. *Resume* has been incorporated into the English language without the two accents.)

Give your "official" name: *Elizabeth Day, William Steven Pattison; not Babbs Day, Billy Pattison.* Middle initials or names are helpful if you have a common name and don't want to run the risk of having your application materials mixed up with those of another applicant with the same or a similar name.

In providing information about your address and telephone number, make sure you can be easily reached at the address and telephone you provide. Some students list campus and home addresses. If you do so, make sure you indicate for how long the campus address will be valid. Some job candidates indicate the hours during which they can be reached by telephone. Others—who do not care that their present employers know they are in the job market—list both work and home telephone numbers. Remember to be complete in providing this information. Completeness here means including area codes for telephone numbers and zip codes for mailing addresses.

Job or Career Objectives

There seems to be some disagreement over the value of including a job or career objective on a resume. In one survey conducted to determine the preferences of recruiting managers in job letters and personal resumes, no mention is even made of the issue.[6]

---

[6] Barron Wells, Nelda Spinks, and Janice Hargrave, "A Survey of the Chief Personnel Officers in the 500 Largest Corporations in the United States to Determine Their Preferences in Job Application Letters and Personal Resumes," *ABCA Bulletin,* 44 (June, 1981), 3–7.

If you decide to include an objective on your resume, use some common sense. "Entry-level position in accounting" cannot be your *career* objective, although it might serve as an immediate *job* objective. A carefully worded objective can show a recruiting manager that you are goal-oriented, but an objective that isn't carefully worded can do you more harm than good and lead to your application being rejected immediately.

If you decide to list an objective, avoid being too vague: *A position with a future in my field*. Equally vague is an objective that specifies only a general field or area of employment. *A position in a law firm* could include that of senior partner or porter or window washer.

Be careful not to indicate that your interest in a company or a field of employment is short-lived. At first glance, the following objective may seem an indication of your willingness to get ahead in the world:

> A position in an accounting firm for several years after which time I plan to return to school to earn a law degree.

However, a recruiting manager reads such a statement as an indication that you won't be working for the company very long. Indicating that your objective is

> A position as an Assistant Buyer for Men's Clothing with Macy's

is acceptable, but not on a mass-produced resume that will also end up in the hands of the recruiting managers at Sears, Saks, and Lord & Taylor.

Use a career or job objective on your resume, then, if you can open a door for yourself by including such a statement. Leave an objective off your resume when such a statement may close doors for you or do you harm.

Education and Training

There is always a danger of underselling and overselling yourself on a resume. We find that many of our students undersell themselves when they list evidence related to their education. Having spent four (and sometimes five or six) years at La Salle, the only entry under education that they include on their resumes is a one-liner:

> B.S., Marketing, La Salle University, 1986.

Those students who oversell themselves in presenting evidence related to their education usually try to lead a busy recruiting manager through their preschool, grade school, high school, and college or university careers. In doing so, they have just lost their reader's interest.

Again, use common sense. If you are 21 or so years old, what difference does it make where you went to grade school? Does where you went to high school even matter anymore? After all, you graduated from high school almost four years ago. A recruiting manager wants to know what you have been doing in those four years *since* high school.

Here is some of the information you can provide a recruiting manager by way of evidence. First, the manager will need to know the name of your school, your degree and major, and your date or anticipated date of graduation. Depending on a variety of circumstances, you may even have to include the name of the city and state where your school is located. Any of the following three examples would provide a recruiting manager with needed information:

M.B.A., (expected) 1988, The Wharton School, University of Pennsylvania; Major: Theatre Arts Administration

B.F.A., 1978, Rhode Island School of Design; Major: Photography

A.S., 1982, Lenoir Community College, Kinston, NC; Major: General Business

Don't overlook the possibility of making your evidence more specific by including information about special features of your educational credentials such as dual or double majors, minors, areas of concentration, or special honors.

Each of the following three examples calls a recruiting manager's attention to an additional piece of evidence in your pool of educational data:

B.S., 1981, Foothill College, Los Altos, CA
Majors: Biology and Chemistry
Minor: Communications

B.B.A., 1985, University of Alabama, Birmingham
Major: Accounting, with strong secondary concentrations in Communications and Computer Science

B.A., *cum laude*, 1986, Old Dominion University, Norfolk, VA
Major: Geography.

If you studied abroad or in a consortium program, say so. If your grade point average or academic honors are impressive, list them.

If your grade point in your major is higher than your overall grade point, just use the grade point in your major. If your grade point last semester was your highest to date, just use the last semester's grade point and label it so. Some candidates list their grade point averages in comparison to the highest grade point average possible at their school.

In listing honors and distinctions, be specific and explain any items that are not self-explanatory:

B.A., with honors, 1985, Marquette University
Major: French; Minor: Computer Science
Major Grade Point Average: 3.8 (out of 4.0)
Junior Year Abroad, International Study Center, Paris
Summer Consortium Study Program, Université de Laval, Quebec, 1980
Outstanding Senior, College of Liberal Arts, 1983
Marcel Orleans Atkins Prize for Outstanding Achievement in the Study of Foreign Languages, 1982

On your resume, you may also want to include a short list of relevant courses you have taken. Remember that recruiting managers are not always specialists in the field in which you are applying. Also, programs in accounting or marketing or psychology differ in their requirements from school to school.

In listing courses, you can include major requirements and electives, especially electives that make you a more attractive job applicant. A student majoring in management might make her resume stand out from the pack with the following lists of courses:

Major Courses	*Elective Courses*
Production Planning and Control	Computer Science
Business Policy	Business Writing
Small Business Consulting	Oral Communications
Organizational Analysis	Economics
	Statistics

Likewise, a student majoring in the humanities or liberal arts might make his resume more impressive with the following lists of courses:

Major Courses	*Elective Courses*
Journalism	Computer Science
Writing for the Media	Accounting
Television Production	Economics
Business Writing	Marketing

When you think you have listed every possible item of interest related to your education and training, check again to make sure you haven't left out any evidence that will make you a more attractive job candidate.

If you indicate you are computer literate, can you actually program in one or more computer languages? What software packages are you familiar with, and what hardware have you used? If you indicate you can read, speak, or write in a language other than English, how well can you do so? Are you really bilingual, or can you only say "the sky is blue" in German and "the pen of my aunt is on the table" in French? Can you operate any special equipment or machinery? Did you have any previous on-the-job training or training in the military that might be of value to a future employer? In short, have you pointed out all the special features of the product, yourself, that you are trying to sell?

Remember that busy recruiting managers are looking for items on a resume that will translate into job skills or job preparation skills. Don't limit your presentation of evidence related to your education and training to one or two lines. Tell a recruiting manager as much about yourself as you can.

Previous Work Experience

In listing your previous work experience, don't be defensive. No recruiting manager expects a soon-to-graduate or a recently graduated college senior to have had experience running a Fortune 500 corporation.

Recruiting managers want first to know that you have worked. Then they want to know where, when, and in what capacity. They want you to indicate that you are

hard working
trustworthy
dependable
capable of working alone
willing to work as a member of a team
punctual
loyal
responsible

We could go on with a list of attributes, but the point we are trying to make is simple. Even the seemingly most menial job—counter person at a fast food restaurant—allows you to demonstrate you have these attributes. Don't apologize for being a gas station attendant. Accentuate the positive. Point out that you worked the graveyard shift or the swing shift or the busiest shift. Point out you had to work with people. Point out that you had to keep accurate records. Point out that you were trusted to keep track of someone else's—your boss's—money.

A recruiting manager on a visit to La Salle once asked a student who "had only worked as a checker in a large supermarket" how much money passed through the student's cash register in a week. The student quickly answered, "$25,000 to 30,000." The recruiting manager was just as quick to point out that "only working as a checker" certainly involved being rather responsible and trustworthy, especially when a student indicated on a resume how much money he or she handled during an average week.

As with the evidence you included under the category of education and training, there is no way to arrange or present evidence under the category of previous work experience. Here, however, are some suggestions to think about:

1. Try to show a pattern in your work experience. Rather than listing every single job or the same job held over successive semesters or summers as separate items, group related evidence in a logical series of entries.
2. Separate out work experience more clearly career related—internships and co-op positions, for instance—from jobs you held just to pay tuition and rent.

3. Don't overlook volunteer work, especially when it is career-related. (Many of La Salle's accounting majors, for example, prepare tax returns for senior citizens and needy families free of charge each year.)

4. Use parallel structure and itemized lists, rather than complete sentences (I did X; then I did Y), to get across as much evidence to a busy recruiting manager as quickly as possible.

5. At the same time, avoid being cryptic. Briefly explain any entry that is not self-explanatory. What sort of company is Biddle, Wharton, Drexel, Jones? Dental, legal, accounting, demolition? Service station attendant and waiter or waitress are self-explanatory, but what's an Assistant to the Trunks Facilities Manager? For that matter, what's a Trunks Facilities Manager? (In case you are wondering, by the way, he or she works neither for the zoo nor for a moving company. A Trunks Facilities Manager oversees telephone lines running throughout a city or state.)

6. Since you are providing details on what you have done, use verbs, action words, at the beginning of each entry detailing your duties or responsibilities. (Use present verbs for information on jobs you still hold and past tense verbs for information on jobs you no longer hold or for activities you have completed.)

Here's a list of action and self-descriptive words and phrases you can use in both a resume and a cover letter. The list is taken from a handout on resumes distributed by La Salle's Career Planning and Placement Bureau:

Action Words

administered	delegated	instituted	re-established
allocated	designed	introduced	regulated
amplified	determined	listed	reinforced
analyzed	developed	maintained	reshaped
arbitrated	devised	managed	restituted
arranged	diagramed	modified	restored
assisted	directed	negotiated	revamped
augmented	distributed	organized	revised
broadened	examined	overhauled	scheduled
catalogued	extended	planned	suggested
calculated	fortified	prepared	supervised
condensed	guided	preserved	systematized
conducted	harmonized	rectified	trained
contrived	headed	recommended	unified
controlled	implemented	recorded	widened
coordinated	initiated	recruited	wrote
created	installed	reduced	

Self-Descriptive Words

active	alert	attentive	consistent
adaptable	ambitious	broad-minded	constructive
aggressive	analytical	conscientious	creative

Self-Descriptive Words *(cont.)*

dependable	extroverted	optimistic	respective
determined	fair	perceptive	self-reliant
diplomatic	forceful	personable	sense of humor
disciplined	imaginative	pleasant	sincere
discrete	independent	positive	sophisticated
economical	logical	practical	systematic
efficient	loyal	productive	tactful
energetic	mature	realistic	talented
enterprising	methodical	reliable	will travel
enthusiastic	objective	resourceful	will relocate

The following are sample entries for possible inclusion on resumes. Again, remember there is no magical, one right way to list evidence. Evidence simply needs to be self-explanatory, complete, and helpful to your advertisement for yourself.

1984–Present
> Maxwell's Drive-In Restaurant, Reno
> *Night Manager*
> Duties: Manage the restaurant and staff from 6 PM until 2 AM. Schedule and oversee staff of 21. Balance registers. Inventory and order food and supplies.

1985–1987
> Ohio Bell Telephone, Columbus Regional Office
> *Customer Service Manager* (1986–1987)
> Duties: Supervised 37 service representatives. Resolved any major complaints on the part of residential or commercial customers. Acted as a liaison with senior management.
> *Customer Service Representative* (1985–1986)
> Duties: Resolved routine complaints and service-related problems for customers. Processed orders for installation, disconnection, and additions to residential service.

Summers 1985–87
> Recreation Park District, New Brunswick, Maine
> *Grounds Keeper and Auxilliary Life Guard*
> Duties: Maintained 150-acre wooded park, two olympic-size swimming pools, 16 tennis courts, and a field house. Took over as Life Guard when needed.

Other Evidence

Finally, take another look at your experience inventory to see if it includes evidence not already listed on your resume. Such evidence might include:

- Articles published in school-sponsored or other publications.
- Talks or speeches given at school-related or other activities.

- Extracurricular activities, including any offices you may have held.
- Activities in professional societies.
- Political activities (don't list party affiliations).
- Church activities (but go easy here).
- Civic or charitable activities.
- Military experience.
- Hobbies, special interests, or skills (but only if you can relate them to the job for which you are applying).

Five Sample Resumes

Format and organization can vary, but make sure you type or have your resumes typeset and duplicated on quality paper. Your resume must make an immediate, good visual impression. Don't pinch pennies. Use a good electric typewriter with a new carbon or film ribbon. If you go to a commercial duplicating or printing shop, choose one that does quality work. You don't have to have your resume printed, but you can't run it off on the nickel or dime copy machine in your school or local library.

Following are five sample resumes for study.[7] Each is chronological in its organization and format. Remember that these are samples, not prescribed models. Be flexible, experiment with format and organization, but put together the best possible advertisement for yourself that you can.

Marylou Hartnoll has yet to graduate. She includes a job objective that is neither too general to mean nothing nor too specific to close doors for her. She clearly points out that her psychology major has made her more attractive as a job candidate by noting her strong additional course work in business. She then expands upon this general information by providing a brief list of courses in her major and in other areas that tells a busy recruiting manager more about herself.

Marylou could have listed her awards and honors next, but she chose instead to discuss her work experience. Notice she has not worked for AT&T or Mobil Oil, and she hasn't been a senior vice-president—no recruiting manager would expect that. Instead, she has worked at the university, in a division related to her job objective. She has also been careful to define terms that may be unfamiliar to a recruiting manager and combined bulleted items in a list with full-length sentences when necessary to provide a maximum amount of information about herself and her job duties.

Ruth Smeltzer spreads information on her resume vertically rather than horizontally as Marylou did. Since most resumes from job candidates still in school or only recently graduated can be expected to fill one full page, you can use a more horizontal format for your evidence when you have a

---

[7] The sample resumes and cover letters used in this chapter were all written by present and former students. Names, addresses, and telephone numbers have been changed. All other data actually appear on the resumes and in the cover letters.

MARYLOU HARTNOLL
843 Northwood Avenue
Elkins Park, PA 19117
(215) 331-0315

JOB OBJECTIVE

To obtain a position in the field of Personnel utilizing the knowledge I have acquired in Psychology and Business.

EDUCATION

La Salle University, Philadelphia, PA

Will obtain a Bachelor of Arts degree in May, 1987.
Major: Psychology, with strong additional course work in Business.
Cumulative Grade Point Average: 3.67/4.00

MAJOR SUBJECTS	ADDITIONAL COURSES
General Psychology	Writing for Business
Personality Dynamics and Adjustment	Accounting Theories, Parts I & II
Developmental Psychology	Micro and Macro Economics
Theories of Personality	Introduction to Sociology
	Philosophy of Work and Culture

WORK EXPERIENCE

Administrative Assistant, Continuing Education for Women (CEW) Program, La Salle University, Philadelphia, PA, from September, 1983, to present. (The Continuing Education for Women Program is designed for women with little or no previous college experience and helps these women in the transition to college.) Duties include:

Assist the Director in general office procedures.
Counsel students in course rostering, study procedures, and personal aspects.
Train work-study students.
Create documents on word processor.
Input student inquiries into computer.
Administer and evaluate testing for prospective students.

I have compiled *The CEW Survival Guide for Continuing Education*, a handbook for new students, and a *Company/Alumni Reference List*, which is used by the Corporate Relations Group at La Salle University.

HONORS AND AWARDS

The Philadelphia City Scholarship: Awarded in 1983, a partial four-year scholarship in recognition of superior academic achievements.

The Greenberg Memorial Scholarship: Awarded in 1984, a partial scholarship for academic excellence.

REFERENCES

References will be furnished upon request.

---

great deal of evidence to present. You can use a more vertical format for your evidence when you have less evidence to present.

The major headings in Ruth's resume run down the left side of the page; those in Marylou's were centered above horizontal blocks of evidence.

Ruth is careful in her objective statement to distinguish an immediate objective from a long-term career. She tells a recruiting manager about what

RUTH SMELTZER
80 DURANT AVENUE
STATEN ISLAND, NY 10306
(718) 572-4366

OBJECTIVE

Challenging entry-level position in Sales or Marketing, with opportunities for promotion to management.

EDUCATION

LA SALLE UNIVERSITY, Philadelphia, PA, May, 1985.
B.S., Business Administration (Marketing major),
GPA: Major 3.75/4.0; Overall 3.5/4.0

MARKETING COURSES	SUPPORTING COURSES
Personal Selling	Computer Science - BASIC
Advertising	Introduction to Information Systems
Marketing Research	Business Writing
Marketing Management	Effective Speech

Honors and Activities: Dean's List, National Business Honor Society, Marketing Association, Intramural Sports.

PENNSYLVANIA STATE UNIVERSITY, Abington, PA
Marketing curriculum, September, 1981, to February, 1983.

BISHOP MCDEVITT HIGH SCHOOL, Wyncote, PA, 1981.
Class ranking: 6/370. National Honor Society; Co-Captain, Marching Unit; Senator, Community Service Corps; Student Council Representative; Debate Team.

EXPERIENCE

LA SALLE UNIVERSITY MBA OFFICE. Administrative Assistant (part-time), 1984-1985 academic year. Aids MBA Program Director in daily administration and coordination. Maintains computer-based directory of program's members.

GULLIFTY'S RESTAURANT, Elkins Park, PA, Summer 1984. As relief hostess, supervised staff. As waitress, handled customer service and menu promotions.

HOT SHOPPES RESTAURANT, Jenkintown, PA, 1981-1983 (part-time). As hostess, coordinated dining room service and supervised ten employees. As waitress principally, handled customer service and menu promotions.

PERSONAL

Financed 60% of my college education.
Interests: cycling, aerobics, jazz dancing, sewing.

REFERENCES

Will be provided upon request.

majoring in marketing at her school requires by listing four major courses she has taken. Realizing that recruiting managers value candidates with some degree of computer literacy and with strong communications skills, she includes a list of supporting courses that demonstrate on paper that she has had training in areas of importance to any future job.

Ruth does include some information about a junior college—a two-year

RUSSELL LEIB
542 Pepper Road
Huntingdon Valley, PA 19006
(215) 632-1970

OBJECTIVE	To obtain a challenging position as a manager in a company with opportunities for advancement.
EDUCATION	**LA SALLE UNIVERSITY, Philadelphia, PA** Bachelor of Science Degree in MANAGEMENT and ACCOUNTING. Graduated *cum laude,* with Honors, in May, 1985. Dean's List for three years. Grade Point Average: 3.5/4.0.
SKILLS	Basic and COBOL Computer Language Statistics
ACTIVITIES	Society for the Advancement of Management Honors Program Business Honor Society Accounting Student Board
EMPLOYMENT	Used to finance 90% of my College Education. **Victoria Station, Inc., Philadelphia, PA** Waiter and Bartender—Provide quality service under time-volume pressures while maintaining good customer relations. Cook—Supervise the preparation and cooking of dinners, maximizing servings from bulk raw products, and maintaining cleanliness to meet the health requirements that apply to restaurants. November, 1980, to present. **GCC Northeast Movie Theatre, Philadelphia, PA** Usher—Checked tickets, maintained order, and supervised over-the-counter sales of refreshments. Summer 1980. **Cool-Rite Appliance Store, Philadelphia, PA** Stockboy—Supervised inventory and distribution of merchandise for pickup and delivery. May, 1979-May, 1980. **Cheese Shop, Neshaminy Mall, Cornwells Heights, PA** Handled sales and food preparation, including sales of specialty and gourmet foods as well as stockwork. Summer 1978.
REFERENCES	Available upon request.

branch of Penn State—and the high school that she attended, but she realizes that a recruiting manager is more interested in information about her present school; therefore, she devotes much more space on her resume to providing information about what she has done at La Salle.

Ruth has held three different jobs. Each is listed with a maximum amount

ANGELA J. HARRIS
745 Hampton Lane
Somerdale, NJ 08083

(609) 884-1395

EDUCATION:	**LA SALLE UNIVERSITY** Philadelphia, Pennsylvania. Bachelor of Arts Degree, May, 1985. **Major: English Concentration: Writing**
	Successfully completed 21 credit hours in writing courses. Major assignments included writing articles, brochures, and advertising copy. Additional courses completed in education, marketing, and BASIC computer programming.
	Financed 95% of education through part-time jobs.
EXPERIENCE:	**Title Clerk. Bill Van's Auto Tags. Roslyn, Pennsylvania.** August, 1983-present. Duties include: transferring automobile titles; issuing temporary license plates, registrations, driver's licenses, and insurance binders; receiving and distributing work from the Department of Transportation.
	Tutor. Jenkintown Grade School. Jenkintown, Pennsylvania. February, 1983-May, 1983. Worked as a volunteer in a fourth-grade classroom assisting students in grammar and writing.
	Office Assistant. Independent Software Affiliates. Jenkintown, Pennsylvania. January, 1982-June, 1982. Duties included: answering and routing phone calls, proofreading and typing software manuals, typing invoices, and updating files.
	Sample Clerk. William A. Popp & Associates. Jenkintown, Pennsylvania. June, 1980-October, 1982. Duties included: designing yarn samples for customers, typing, filing, and checking credit records.
SPECIAL ABILITIES:	—Type 80 W.P.M. —Shorthand —Computer literate —Notary Public

REFERENCES AND WRITING SAMPLES AVAILABLE UPON REQUEST.

of information. Ruth also indicates that she has participated in extracurricular activities. A recruiting manager is also likely to be impressed with the fact that she mentions she has financed 60% of her college education.

Russell Lieb's resume uses all caps to highlight the fact that he has a double major. While he does not list specific courses he has taken, he does

JANE MARIE BIDDLE
2741 Fairway
Baltimore, MD 21222
(301) 338-3253

EDUCATION:
9/82-5/86

LA SALLE UNIVERSITY   Philadelphia, PA
Bachelor of Science Degree in Business Administration, May, 1986.
Dual Majors: Accounting and Management Information Systems
Cumulative Grade Point Average: 3.4 (out of 4.0)
Accounting Grade Point Average: 3.4 (out of 4.0)

HONORS AND ORGANIZATIONS

- Dean's List throughout college career.
- Elected class representative for Executive Student Accounting Board.
- Awarded W.W. Smith Charitable Trust Grant for academic achievement.
- Beta Alpha Accounting Honors Society member.
- Business Honor Society member.
- Junior Achievement advisor.
- Women's Crew team.

1/85-5/85

TRINITY COLLEGE DUBLIN   Dublin, Ireland
Study abroad program through Temple University.
Concentration: Irish literature, history, and culture.

3/84-6/84

ADELPHIA REAL ESTATE SCHOOL   Philadelphia, PA
Concentration: Real Estate practices and fundamentals.

EXPERIENCE:

Financed 50% of my college education through grants, loans, and the
following jobs:

6/85-present

CIGNA CORPORATION—INA REINSURANCE CO.   Philadelphia, PA
Underwriting Assistant in International Facultative Department.
- Calculate liability authorizations and compensating premiums considering
  multiple and varying deductions. Task requires considerable attention to
  contract terms and an ongoing awareness of regional statutory requirements.
- Construct a data base management system for the Claims Department
  utilizing the software package DATAEASE.
- Devise numerous accounting worksheets using LOTUS 1-2-3 to record and
  monitor claims activity and pricing variances.
- Created a computer program to generate account statements and to assist in
  reconciling accounting discrepancies.
- Drafted a Master Contract Information form after consultation with the
  Assistant Vice-President of International Reinsurance. Project laid the
  foundation for the Facultative Accounting and Statistical System.

include an entry about possible career-related skills he has acquired. His work experience is somewhat more extensive than those of Marylou or Ruth, but again that experience is not directly career-related. Russell does, however, tell a busy recruiting manager a great deal of information about his work experience that the manager can easily translate into useful job

7/84-12/84	**E.I. DU PONT DE NEMOURS & CO., INC.** Wilmington, DE

Junior Cost Accountant in Comptroller's Department.
- Analyzed and reported monthly production costs for two manufacturing plants.
- Streamlined monthly and annual manufacturing reports using LOTUS 1-2-3 on an IBM PC-XT. Project resulted in an 80% time savings and arithmetical accuracy.
- Developed a long-range construction cost forecast using an on-line software package.
- Participated actively in weekly staff meetings on quality control and short-term planning.

9/83-6/84	**A.B. REALTY CO., INC.** Philadelphia, PA

Property Manager and Bookkeeper.
- Maintained all receipt and disbursement journals accurately and completely.
- Recorded rental income from operations.
- Prepared and published monthly property management reports for investors.
- Provided sales and financial information to clientele.

9/83-5/84	**LA SALLE UNIVERSITY** Philadelphia, PA

Accounting Tutor in the Academic Discovery Program.
- Instructed and reinforced accounting principles for accounting students.
- Reviewed and improved effective study skills.
- Emphasized test preparation skills to improve exam results.

8/82-8/83	**PENNSYLVANIA ASSOCIATION OF STUDENT COUNCILS (PASC)** Shippensburg, PA

Group Leader and Special Interest Seminar Director.
- Selected by the PASC Executive Board to conduct training workshop for annual leadership conference.
- Designed and directed communication and assertiveness seminars.
- Motivated and supervised high school delegates toward group goals.

5/82-9/83	**OFF-THE-RAX** Philadelphia, PA

Sales and Stock Associate.
- Performed cash register operations for a high volume clothing store.
- Managed extensive merchandising and stylizing of stock.
- Conducted inventories and handled price revisions.

REFERENCES:	Available upon request.

skills for a position in management or accounting. Like Ruth, Russell points out that he has financed a part—in his case, the major part—of his college education. This simple piece of evidence speaks volumes to a recruiting manager.

Originally, Angela J. Harris thought she was "just an English major" who had little to offer a potential employer. As you can see from her resume, though, she does have much to offer a potential employer once she carefully assessed what she has learned in all those English courses and what skills she has picked up from her various part-time jobs. No one will hire Angela as an accountant or a computer programmer—she wouldn't want either job in the first place. But Angela's resume makes her an attractive candidate for a variety of positions in any number of businesses and industries. These positions, by the way, are not simply secretarial. A bank interested in a customer relations expert or a computer firm looking for a technical writer would find Angela an attractive candidate easily trained to be a productive and valuable employee.

Jane Marie Biddle is an exception to a general rule—she needs more than one page for her resume. Notice, however, that her resume is not padded. Jane simply has managed to accumulate a wealth of work experience, while also attending school full-time, participating in extracurricular acties, and winning recognition for her academic work.

Jane isn't Wonder Woman. She has, however, been fortunate enough— and perhaps foresightful enough—to prepare herself for a career in business and industry as thoroughly as possible. Even given all this preparation, she isn't yet sure what her career or job objective is. Consequently, she omits any mention of an objective on her resume. She is willing to go "on the job market" to see what prospects await her. She will be especially attractive to recruiting managers looking for job candidates with the skills her various jobs and her education have provided her with. In this respect, she is no different from Marylou, Ruth, or Russell. All four students have written excellent advertisements for themselves. Their resumes differ for the simple reason that no two job candidates—even for the same position— are ever the same.

Functional Resumes

The sample resumes we just presented are traditional or chronological resumes. Such resumes are not ideal ways for all candidates to present the evidence they have to offer a recruiting manager.

Chronological resumes are ideal for job candidates who have experience to play upon. Other candidates—students majoring in nonbusiness or nontechnical areas, candidates returning to work after an extended absence from the world of work, or people interested in changing careers—might benefit more from a functional resume.

A functional resume provides you with some flexibility in emphasizing your evidence. It allows you to de-emphasize experience while emphasizing the abilities and skills you would bring to a job.

Your purpose in submitting a functional resume is still to get an interview, but you may have a more difficult time making an impression on your audience. Here's a functional resume put together by an English major seeking an entry-level position with an insurance company in the company's policy holders' service department. The applicant has highlighted his abilities and skills in areas that he hopes will impress the recruiting manager.

Qualifications of

KEITH ARUTURIAN

for a Management Trainee Position with
the Whole Life Insurance Company

CAREER OBJECTIVE: A non-sales career position with an insurance company.

Functional Review of Previous Experience

**EXCELLENT
COMMUNICATION SKILLS:** B.A. in English, La Salle University, 1984
University student newspaper:
   Editor-in-Chief, 1983-84
   Sports Editor, 1982-83
   Reporter, 1981-82

**PROVEN MANAGEMENT
SKILLS:**   As Editor-in-Chief, supervised a staff of 35, handled all dealings with printers and vendors, managed a $65,000 budget

**PROVEN LEADERSHIP
SKILLS:**   Elected three times as Senator to University Student Senate
Appointed Alternate Student Representative to the University
   Board of Trustees

**HARD WORKING AND
DEPENDABLE:**   Paid for half of all of my college expenses by working part-time and
   summers as a waiter at Fred and Jerry's Restaurant; despite heavy
   work schedule, maintained a 2.9 grade point average (out
   of possible 4.0).

COMMUNITY MINDED: Chairman, Annual Toys for Tots Drive, University Student Senate, 1984
Organized a Blood Donor Drive at college and at work, 1983

Address and Telephone Numbers

56 North Olney Avenue   215-555-9999 (days)
Philadelphia, PA 19141   215-444-8867 (after 5 p.m.)

Cover Letters: Evidence

Since not every job you apply for will require you to submit a cover letter, you will probably submit more resumes than cover letters. A cover letter repeats some of the evidence contained in your resume, but it allows you to target your evidence more directly to the needs of a specific recruiting manager or to the interests of a specific company. Generally, you send out the same resume each time you apply for a job—though some candidates may have two or even three different resumes organizing the same evidence with a different emphasis. Cover letters are always individualized to meet the demands of a specific job.

A cover letter is simply a type of business letter. When you actually compose the letter, review our comments on possible formats in Chapter 4. Do not, however, use letterhead from your present employer—you are stealing company property if you do. Simply use plain bond, and include your return address directly above and aligned with the date of your letter.

There is no great secret to compiling evidence for a cover letter. A recruiting manager will want to know the following:

- The job you are applying for.
- Where you found out about the job vacancy.
- Your credentials for the job.
- Which particular items listed on your resume are especially useful to you if you should be fortunate enough to be offered the job.
- Whether you're available for an interview or are prepared to submit additional application materials (a portfolio or writing samples, for instance).

If you are responding to a printed advertisement, pick up on the key words and phrases in that advertisement when you present your evidence. Indeed, your evidence and the specifics in the advertisement should parallel one another as closely as possible. If the advertisement calls for someone with a degree in management, make sure you mention that you have such a degree early in your letter.

If you are applying for a job that is unannounced or unadvertised, don't waste your busy reader's time. State specifically what job or area you are interested in and what you have to offer an employer.

Think of your cover letter as your one chance to make a specific sales pitch for a specific job with a specific company. Then emphasize specific pieces of evidence that directly relate to the company's needs and interests. Show a recruiting manager that you know something about his or her company and that you know how you can relate what you know about that company to what you can offer that company.

Our earlier suggestions about how to present evidence on a resume apply just as well to cover letters. Be positive, be assertive—not aggressive, just assertive—be clear, explaining anything not self-explanatory, and, above all, be specific. Nothing prompts a quicker rejection than a vaguely worded cover letter.

Three Sample Cover Letters

Following are three sample cover letters; again, they are not meant to be absolute models. They simply suggest approaches to different job application situations.

In the first letter, the applicant is responding to the following newspaper advertisement:

EDITOR/PHOTOGRAPHER

If you have a degree in Journalism, English, Communication, or the equivalent, with some on-the-job experience writing for in-house publications . . . and if you are able to demonstrate news judgment plus writing, editing, layout, and photography skills, we have an excellent career opportunity for you in our Employee Communication Department.

Forward your resume to Ms. Tina Finkelstein, McNaughton Pharmaceuticals, Spring House, PA 19476.

```
                                        437 Royal Street
                                        New Orleans, Louisiana 70130
                                        December 18, 1986


Ms. Tina Finkelstein
McNaughton Pharmaceuticals
Spring House, PA 19476

Dear Ms. Finkelstein:

I wish to apply for the position of Editor/Photographer
listed in yesterday's Chicago Tribune. I hold a bachelor's
degree in English from Tulane University, and I have been
employed by Loyola University in New Orleans as assistant
editor and staff photographer for all in-house
publications for the last 9 months.

I am especially interested in the broader scope of duties
the position with McNaughton offers. I would bring to
these duties experience in all aspects of the publication
of employee-directed communications.

My training at Tulane included a strong minor in
photography and a secondary concentration in employee
relations. While I was an undergraduate, my photographic
```

spreads were three times honored by the Association of College and University Newspapers.

Most recently, I received honorable mention for a photographic study of New Orleans entered in a National Geographic Magazine contest for layouts on American cities.

I would be happy to provide you with samples of my work, and I will be in your area during the second and third weeks of January to attend various activities of the National Association of Jesuit Colleges and Universities which is meeting in New York. I would welcome an opportunity to speak with you personally about this position.

I have enclosed a copy of my resume, and I would be more than happy to provide you with any additional information you might need. I look forward to hearing from you further about what sounds like an exciting and challenging position.

Sincerely,

Arthurina K. Washington

Arthurina K. Washington

Encl.

The second letter is in response to a blind advertisement in a newspaper. The advertisement provides only sketchy information about the job and its requirements. Such an advertisement is difficult to respond to, since you run the danger of overselling or underselling yourself. Here's the advertisement and the letter:

PSYCHIATRIC AIDE. Full time to assist mental health professional with psychiatric clients in MH/MR center in the District. Some college or a degree plus 1–3 years experience with psychiatric clients required. Reply Box 897-J, The Post.

189 Rhode Island Avenue NW
Washington, DC
May 19, 1986

Box 897-J
The Washington Post
Washington, DC 20030

I wish to apply for the position of Psychiatric Aide advertised in today's Post. I will graduate next week from

American University with a bachelor's degree in Psychology
from a five-year co-op program. As a member of that
program, I spent a year working full-time at St.
Elizabeth's Hospital. In addition, I have worked part-time
at nights in the Psychiatric Ward at Georgetown Hospital
as an Orderly and Patient Supervisor.

My studies at American University included not only a
major in psychology but minors in biology and chemistry.
My senior seminar paper was a study of outpatient
psychiatric facilities in the District.

At St. Elizabeth's and Georgetown Hospitals, I worked
closely with doctors and other staff in treatment and
therapy programs for patients with varying degrees of
mental disability. At St. Elizabeth's, I also studied art
therapy in the hospital's new pilot program, where my
fluency in Spanish proved an asset.

I have enclosed a copy of my resume, and I would be happy
to provide you with letters of reference. I look forward
to hearing from you about the opening on your center's
staff.

                              Sincerely,

                              José A. Hornedo

                              José A. Hornedo

Encl.

The second letter strikes a balance between overselling and underselling.
Obviously, it would be easier to write for the job if you knew the name
of the center that had run the advertisement and could do some homework
before writing the letter.

The next letter is a prospecting letter. The applicant is writing to a company
she'd like to work for, but the company hasn't run an advertisement announcing any specific vacancy. In a prospecting letter, you need to find out all
you can ahead of time about the company you are writing to, and then
you need to direct your appeal for a job to the company's current needs
and interests. For a prospecting letter, never use a blind address. Find out
the name of the person in charge of hiring, and address your letter to that
person.

Here's a sample. Note how the applicant tries to show that she can be
both specialist and generalist.

2520 Cimmaron Street
Los Angeles, CA 90018
November 18, 1986

Mr. Devin Adair
Director of Personnel
Cummings Financial Services
2727 Sand Hill Road
Menlo Park, CA 94025

Dear Mr. Adair:

I will graduate next month from UCLA with a bachelor's
degree in business administration. My major is accounting,
and my double minor is in finance and English. While at
UCLA, I interned at Bache and Company, and I worked
seasonally for H & R Block Tax Services.

I wish to start a career in financial services after
graduation, and I can offer your company the specialized
skills of an accountant and the general business and
communications skills that my minors afforded. My
internship and work experience allow me to offer your
company in addition a working knowledge of the stock and
commodity exchanges as well as actual on-the-job
experience in public accounting.

I am a hard worker--my grade point average has remained
above 3.0 (out of a possible 4.0)--and I have worked
throughout college to pay for my own education.

I have enclosed a copy of my resume, and I would like an
opportunity to talk to you in person about what I can
offer your company as it expands its market in southern
California. I will call your office to schedule an
interview after the Thanksgiving holiday.

Sincerely,

Connie Y. Chen

Connie Y. Chen

Follow-Up Correspondence

If you are invited to an interview, confirm the time and place of the interview
in writing. If you have had an interview, immediately write to the person
who interviewed you to say thank you and to restate your interest in the

position. If there is a sudden addition to your credentials, inform whomever you made your original application to in a brief note. Type all such notes. Here are three sample follow-up notes; notice the brevity of each:

Dear Mr. Haight:

Thank you for your telephone call this morning. I look forward to meeting you for an interview next Tuesday, September 9, at 10:00 A.M. in your office.

                                                  Sincerely,

Dear Ms. Silva:

Thank your for the opportunity to talk about the position with your company. I am still very much interested in the vacancy in your accounting department. Given my double major in computer science and accounting, I would welcome the opportunity to work with your staff now that they are automating all your accounts.

                                                  Sincerely,

Dear Mrs. Davis:

Since writing to you last week to apply for the internship in the Accounting Department, I have been elected to Phi Beta Kappa. I am still very interested in this internship, and I look forward to hearing from you about my application soon.

                                                  Sincerely,

Letters Refusing Job Offers and Resigning From Positions

Whenever you turn down a job offer or resign from a position, no matter what your reason, do so in a short, polite, no-nonsense letter. When Richard Nixon resigned the presidency of the United States, he simply wrote the following one-sentence letter to then-Secretary of State Henry Kissinger:

Dear Mr. Secretary:

Effective at noon today, I resign the office of President
of the United States.

                                              Sincerely,

You don't need to say anything more. You never know when you may want
a job back or when you may want to reapply to a company. Simply say
"no" or "goodbye," and leave your letter at that.

Exercises

1. In writing a cover letter, you have to steer a course between conceit and legitimate
   self-promotion.

   Here's a memo written in 1917 by a junior officer in the United States Army.
   The officer is applying for a position in the then-new service of tanks. The memo
   was forwarded through channels to the commander-in-chief of the American forces
   in Europe. The subject line read "Command in the Tank Service."

   How well does this memo steer the course between conceit and legitimate
   self-promotion?

   (To allow you to assess the effectiveness of this memo objectively, we have
   delayed indicating the source of this "cover letter" until the end of all the exercises
   in this chapter.)

   I understand that there is to be a new service of
   "Tanks" organized and request that my name be considered
   for command in that service. I think myself qualified for
   this service for the following reasons. The duty of
   "tanks" and more especially of "Light Tanks" is analogous
   to the duty performed by cavalry in normal wars. I am a
   cavalryman. I have commanded a Machine Gun Troop and know
   something of the mechanism of machine guns. I have always
   had a Troop which shot well so think that I am a good
   instructor in fire. It is stated that accurate fire is
   very necessary to good use of tanks. I have run Gas
   Engines since 1917 and have used and repaired Gas
   Automobiles since 1905. I speak and read French better
   than 95 percent of American Officers so could get
   information from the French Direct. I have also been to
   school in France and have always gotten on well with
   frenchmen. I believe that I have quick judgment and that I
   am willing to take chances. Also I have always believed in
   getting close to the enemy and have taught this for two
   years at the Mounted Service School where I had success in

arousing the aggressive spirit in the students. I believe
that I am the only American who has ever made an attack in
a motor vehicle. This request is not made because I
dislike my present duty or am desirous of evading it but
because I believe when we get "Tanks" I would be able to
do good service in them.

2. The following letter was actually sent by a soon-to-graduate student to a major
financial institution. The student attended one of the top three business schools
in the country. His credentials were impressive. The letter was a follow-up to
earlier correspondence with the financial institution. Comment on the letter's
strengths and weaknesses.

Dear Mrs. Kiel:

  This May, I will graduate from State University with a
B.A. in Economics. I am interested in securing a position
as a research assistant with your company upon graduation.

  If you recall, you will remember that, upon the
recommendation of an acquaitance of mine, Mike Biddle, I
sent you a resume two summers ago in reference to
obtaining an internship at your company the spring
semester of the past academic year. You replied that you
were impressed with my resume but that no paying positions
as such existed for undergraduates.

  I would like to bring you up to date on what I've done
since then. Following a semester of study abroad, last
spring and continuing through the end of the summer, I had
the fortunate opportunity to work as a research assistant
on a State University-sponsored project concerned with
developing a regional data base and econometric model for
the Sovier Union. Meanwhile, completing my senior year, I
am working for Dr. Woodrow Pate at State.

  Enclosed is a copy of my updated resume highlighting
pertinent aspects of my experience. My course work in
Economics, Mathematics, Regional Science, and Computer
Science has equipped me with basic skills necessary for
economic analysis. In addition, I've gained valuable
experience over the past three years through the various
research assistantships I've held. I believe that my
combined academic and work experience would prove to be
most useful in the position of research assistant at your
company.

  I would appreciate an opportunity to discuss my
qualifications with you in further detail. I will call you

office next Friday to arrange a mutually convenient date
for a personal interview.

Thank you for your time and consideration.

Sincerely,

Bruce M. Mullins

Bruce M. Mullins

3. Here are some more examples of "suicide by letter bomb." Each of the following
   passages is excerpted from an actual cover or application letter. These letters
   were written by students and adults, all intelligent people, all really interested
   in securing the positions they were applying for. Comment on why each will
   ensure an immediate rejection of the applicant.

   a. I am applying for position openings in the Marketing Department of your
      company. From the ad in the Collegiate Career Woman, Vol. 8, # 8, Winter,
      1986/87 magazine, I learned that you have several positions available and I
      feel that I could be the person that you are looking for. I would take any of
      these positions.

   b. You may ask why I want to make this change, only because I am looking
      for advancement and a more challenging and rewarding career. I would like
      to emphasize my experience and my academic training which would be a
      very valuable asset to your organization.

   c. Air travel has always fascinated me. One yar of living abroad in the Republic
      of Panama in 1978 was a great experience in learning about other people's
      customs and life styles. I still have many acquaintances I made in Panama.
      The experience taught me to feel for other's needs. The Panamanian people
      were very hostile toward Americans until they felt you were sincere in your
      feelings toward them.

   d. I am available at your convenience for an interview and would like to be
      considered as a candidate in the above mentioned position.

   e. Although I do not have direct operations experience, I have worked closely
      with operations during audits and during the many conversions that have
      been my responsibility. Even though we are separated by a great distance, I
      am more than willing to attend an interview at your convenience. I can fly
      from here to Denver, and perhaps I could get in some skiing on the side at
      the same time. I realize that my current salary of $42,000 is inflated becuase
      I live in the congested Northeast section of the country. Because I would
      like to work and play in the open spaces of the Rockies, I am willing to
      negotiate a 10 per cent reduction in salary.

   f. At my present position at the West Orange Police Department, I am resposnible
      for keeping the peace at certain businesses. I also write reports and traffic
      tickets. My position as dispatcher will help me do the job better. I understand
      what the dispatcher has to contend with. Furthermore, I know the Milburn
      area well because of my job with Simone's Towing Service.

   g. Writing has been a major part of my professional life for the past twelve
      years and my persistence and ability to follow through has served me well

in my extensive dealing with the press and broadcast media. My present position as public relations director for an architectural firm has prepared me to take brochures, newsletters, and promotional material from idea stage to final project.

h. Finally, the enclosed resume will provide you with information which indicates that I have had much prior experience in being of service to the more affluent people in life. This experience is demonstrated in the form of a highly personable, well-dressed young executive who feels he can be an asset to your firm.

i. I am a fluent speaker, reader, and writer of Spanish. I have four years of helping people in a service capacity under my belt and as a result I can deal well with people even under stressful conditions. I have an intense desire to help people with their emotional problems and own the intuitive skills to be successful at it.

j. As you'll see from my resume, I have extensive experience in nearly all the areas you list as job requirements. If you're interested in getting acquainted, you can reach me at 555-1432 during the day or at my home in the evening, 555-9652. Your response to my inquiry will be greatly appreciated.

k. I am writing to apply for the position of Personnel Manager which you advertised for in the February fifteenth issue of the Cleveland Plain Dealer. As the attached resume indicates, I have extensive experience in personnel, training and employment, as well as a strong management backround.

l. Moreover, every employer I have had has been more than happy to have hired me.

m. In the matter of salary, keeping in mind that you are located in the South, I find a starting salary of $20,500 an average salary for comparable positions.

n. Your ad has impressed me greatly. My ability to move upward thru the ranks of any company that I have worked for assures me that I have demonstrated strong leadership ability and initiative. Presently, I am the Manager of the Settlement Account at my employer. This handles 450 large corporate accounts. Interfacing with these clients and local business organizations is a very vital part of my job today.

o. I have an excellent sales record in the non-profit health insurance field which a sales representative and instituted and structured sales training program for the sales representatives under my management.

4. In preparation for the next exercise, write a job description for a job you now hold. If you are not presently working, write the description for a job you held in the past.

5. Find an advertisement for a job that you would like to have and that you have the credentials to land. Write a cover letter and a resume applying for that job. Then write a letter to a third person asking for a letter of recommendation for the job. Include a copy of the job advertisement as part of this assignment.

(The aggressive applicant who wrote the memo in Exercise 1 on pages 142–143 was none other than George S. Patton. From *The Patton Papers, 1885–1940*, edited by Martin Blumenson. Copyright © 1972 by Martin Blumenson. Reprinted by permission of Houghton Mifflin Company.)

CHAPTER 7

Messages That Inform

There are literally hundreds of kinds of letters, memos, and other messages that fall into the category of informative writing. You write to place an order, to answer an inquiry about an employee or a product, or to congratulate a co-worker on a job well done. You write to inform a supplier that an adjustment is in order, or to give the reader good news or bad news. On occasion, you write to say "thank you" or to express your sympathy. The list of possibilities is extensive, and no simple formula can provide you with instant answers or all-purpose, fill-in-the-blank models. (On a recent visit to a computer store, however, we did see a program that consisted of more than 200 business letters, blanks to be filled in by the sender!)

On the assumption that neither you nor your reader would be happy with such prefabricated letters for all occasions, we again offer the PAFEO process as a reasonable, systematic way of writing an informative message, no matter what specific form it may take.

PAFEO will help you direct your message to the reader so that your purpose is accomplished. Too often informative messages fail to communicate effectively because the writer assumes that the reader will know what is meant. The message is writer-based when it is written, and the writer does not recognize the need to turn it into reader-based prose, using the PAFEO process as a guide.

Informative Writing: Examples for Discussion

To show you how easy it is to have even a short message miss connecting with the reader, here are three examples. For the present, we'd just like you to read them and perhaps discuss them a bit with your classmates.

EXAMPLE 1

SUPPLEMENTAL SECURITY
Income—Notice of Planned Action
Date: 5/30/76

Your payments—or those of the individual named above—will be change as follows:

Your checks will stop July 1976.

Beginning July 1976 your gold-colored check will be stopped.

The law provides for an increase in the federal Supplemental Security Income payment rate beginning July 1976 and in the amount of Social Security benefits beginning July 1976.

Although Supplemental Security income payments are made monthly, we average your income over a calendar quarter in figuring the amount of your payment. We find that for the calendar quarter Jul 1976 through Sep 1976 your average monthly income is $187.80.

Your average monthly income is based on your income for the calendar quarter as follows:

Your Social Security benefit, before any deduction for Medicare Medical Insurance premiums, of $563.40.

An agency of your state will inform you of any change in your eligibility for Medical Assistance—Medicaid—caused by this action.

Although we plan to take the action shown above, you may have your prior payment continued or reinstated if you request an appeal within 10 days of receiving this notice.

If you had a hard time reconciling this message with its intended purpose, think what the actual recipients, most of them women over 62 and men over 65, must have thought. The opening sentences seem designed to induce coronary arrest. The checks will stop; no more money will be coming in. The closing paragraph only reinforces the idea that this message may be bringing bad news. What person making only $187.80 a month would appeal an increase in benefits?

The second message is from an employer to a college senior. The student has done well in a first interview on campus. The employer now wants to invite the student to come to the company for a second interview.

Do you think the organization of the letter makes the purpose clear to the reader from the beginning?

EXAMPLE 2
_____

May 15, 1986

Miss Lynne Clurman
842 Mitchell Avenue
Morton, PA 19178

Dear Miss Clurman:

Thank you for taking the time last week to meet with Toby Nelson, our recruiter, when he visited your school. Toby has forwarded your resume along with his notes from the interview to our hiring committee for review.

I want you to know that the committee was very impressed with your credentials. Your balance of course work and work experience is very impressive.

We would like, therefore, to invite you to visit our company for a series of interviews with members of our staff. We are scheduling these day-long series of interviews during the second and third weeks of June. Please call me at 978-9007 to schedule a day that you would find convenient.

Sincerely,

K. George Trimmer

K. George Trimmer

The message is a good news message, but the opening could be misleading. If you have ever applied for a job and been rejected, you may have received a letter with an opening buffer designed to let you down nicely. A number of our students received variations on this letter after their on-campus interviews. To a person, they at first thought the letter was a rejection. Only by reading the entire letter through did they determine that the letter was really bringing them good news. We only hope none of our other students read only as far as the first paragraph and then threw the letter away.

The third message is a bad news message. One of the alternative long-distance telephone companies writes to let customers know that rates will increase.

Although the announcement of the rate increase is the primary purpose, the company also wants to take some of the sting out of the bad news by pointing out that customers are still saving money on long distance calls. Is the organization of the message effective in achieving both purposes? Do you think the letter does a good job with public relations?

EXAMPLE 3

NOTICE TO CUSTOMERS

Effective July 1, 1986, the prices for evening and night & weekend calls made through our network are increaing 5%.

Evening rates for calls made under 22 miles and night & weekend rates for calls under 55 miles are not affected. Daytime rates and calls to Canada will also continue to remain unchanged.

Even with these slight changes, you will continue to save from 5 to 50% using our network for all long distance calls!

The message does inform. It gives the reader the main piece of information in the first sentence. Unfortunately, the letter is a poor exercise in public relations. The second and third paragraphs do offer the reader something in return for the increase in rates: other rates remain the same, and all rates still save users money. These two points may counterbalance any impulse on the reader's part to cancel service. Since these last two points are stronger from the point of view of public relations, they might make an effective buffer up front to the bad news that there is to be a rate increase.

In all three messages, you get a sense that poor planning has resulted in poor writing. If the writers of these three messages had used the PAFEO process, they might well have written different, more effective communications.

Purpose and Audience

Taking a few minutes to clarify your purpose and think about your audience's needs will pay off in time saved in the long run. By doing so, you may avoid a series of telephone calls from confused (and often irate) readers of your message. Don't settle for a simple answer to the question, "Why am I writing this?" You are writing to inform. True. But what else do you have to consider if you want to write the most effective message you can? Don't fail to recognize *operational purposes* as well as your *general aim*.

Let's illustrate how to recognize operational purposes as well as general aims by looking at Example 3 again (on page 150).

This brief message is actually quite complex when you consider the mixed purposes it must serve and the mixed audiences who will read it, each audience having differing needs to be met. An analysis of the operational purposes as related to the audiences might generate a list of purposes such as this one:

1. To inform the customer of the bad news—the rate increase.
2. To explain the reasons that made the increase necessary.
3. To persuade the customer that the company still offers good value in relation to the competition.
4. To meet any legal requirements of a state regulatory agency governing announcements of rate changes.
5. To satisfy the preferences of the writer's supervisor (and initial reader).
6. To provide a clear factual record for the files.[1]

Format

Your choice for format may be limited only by your imagination, or it may be dictated by established company practice. In either case, let intelligence and your desire to help your readers understand your purpose guide you in your choices. Put what you have learned in Chapter 4 to work. Pay careful attention to the way you arrange your materials on the page. Your information must be accessible to your readers.

Reminder: You can insure accessibility by generously using

Capitalization
Headings
Enumeration
Lists
White space

Generally speaking, it is better to use shorter paragraphs than long ones. Shorter paragraphs invite the reader's eye to continue; long blocks of print are unappealing to the eye. A clear format promises what every reader wants: a clear message.

Evidence

Be sure to give your readers the evidence they need to understand your message. What you want to tell your readers and what they want or need

---

[1] These operational purposes are adapted from our discussion in Chapter 2 and the treatment of mixed purposes and audiences in C. H. Knoblauch, "Intentionality in the Writing Process," *College Composition and Communication,* 31 (May, 1980), 153–159.

to hear may be two different things. You may have spent days figuring our what caused a problem for a customer, but your customer doesn't necessarily need a step-by-step account of how you solved the problem. Your customer may simply need to know that the problem has been solved and that future problems can be avoided in such-and-such a way.

Organization

Your can organize informative writing in a variety of ways. Your choice will depend largely on your purpose and the impact your message will have on your audience. The danger to watch out for is assuming that the organizational pattern of your own thought processes in writing will match the organizational pattern of your audience's thought processes in reading.

You may have solved a problem by studying its background. Your message needs to be organized, however, to define the problem, and perhaps even detail its solution, before you give your readers that background. Your approach to a problem may have been to trace an issue chronologically or spatially, but your readers may be better served by a letter, memo, or report that distinguishes cause from effect or that compares, contrasts, or classifies.

Your organizational pattern must then be based in your reader's experience. You need to transform writer-based prose into reader-based prose and writer-based organizational patterns into reader-based ones. Again, as Linda Flower points out, you can undertake this transformation by setting up a shared goal with your reader, developing a reader-based structure, and giving your reader cues throughout your message.[2]

Organizing Good News Messages

Good news messages are messages that your audience will respond to positively or neutrally. Typically, they might include answers to inquiries, adjustment letters, order acknowledgments, or congratulatory messages.

Since your audience is neutral or sympathetic to your message, your best approach is to be direct. Put the good news right at the beginning and say it as positively as you can. Richard Lanham has good advice: "Start fast—no mindless introductions."[3] Avoid openers like "Your letter was passed on to me for handling by Ms. Lottier, our Director of Marketing."

The following letter demonstrates how to give the reader the most important information first and then pass along additional information that will build goodwill.

[2] Linda Flower, *Problem-Solving Strategies for Writing,* 2nd ed. (San Diego: Harcourt Brace Jovanovich, 1985), pp. 162–173.
[3] Richard A. Lanham, *Revising Prose* (New York: Scribner's, 1979), p. 6.

IMPORTANT . . .
PLEASE READ THIS LETTER

Dear Member:

Enclosed is your new life insurance certificate. This reflects the additional coverage you recently requested.

We are pleased that you share the enthusiasm for this fine group insurance plan with many other members. Please keep this certificate in a safe place with your other valuables.

The certificate indicates the effective date of coverage. The semi-annual billing dates of this plan are June 1st and December 1st of each year. You will receive a premium notice from us prior to these dates.

As your Insurance Administrator, we are dedicated to give you the best service possible. Please direct all correspondence and report all claims to this office so that you may receive prompt and personal attention to every detail of your insurance coverage.

Thank you so much for this opportunity to serve you.

Sincerely,

Had the additional information come first, the writer would be ignoring the reader's need to be informed. The writer has correctly subordinated the company's desire to build good will to the reader's more immediate need to be informed.

If you are writing a memo communicating good news, you have two opportunities to get your main idea out front: the subject line and the first sentence. Avoid vague subject lines like "Smith Account." Tell your reader something about the Smith account in your subject line, and expand upon your subject line in your first sentence. There is a mistaken notion among many business people that subject lines should be only two or three words long. Subject lines should never run to a complete paragraph, but they must give readers some useful information.

TO: Earl Stevens
FROM: Bob Hogan
DATE: March 2, 1986
SUBJECT: Cost Overruns in the Smith Account

For the sixth month in a row, there have been substantial cost overruns in the Smith account. This problem needs your immediate attention.

In longer good news messages, put your main idea in your first sentence, and then use the balance of your first paragraph to preview for the reader what is to follow and how it will follow. If, for instance, you will go on to explain three items related to your main idea, let your reader know so in your first paragraph.

In writing a good news message, you can take some tips from the field of journalism. In your first sentence (for shorter messages) or your first paragraph (for longer messages), try to answer the questions: Who? What? When? Where? Why? and How? But be careful not to end up with an overly long opening sentence or paragraph.

While it is time-honored tradition in business and industry to begin letters and memos with an acknowledgment that previous correspondence has been received, such an opening is superfluous and wastes the reader's time. You can't respond to an earlier message not received. A second time-honored tradition in business and industry is the routing of letters and memos to others for responses. There is nothing wrong with this tradition, but it should be obvious that you are responding to the original correspondence if your name is on the response. There is nothing wrong with expressing thanks to a customer, but such gratitude isn't your main point in a good news message and ought to come last—after you give answers to the questions your reader has raised.

In the middle section of your message, explain your main idea in detail, and break the points of your explanation up so that your reader can easily understand them. If, for instance, you are explaining three subpoints related to your main idea, break your explanation up into three sections clearly identified by headings so that the separate sections of your explanation and the overall organization of your message can clearly be seen by your reader. As you continue through your explanation, make sure that any terms you use or points you make will be readily understood by your reader.

Organize individual sections of your good news messages in order of decreasing importance. In doing so, you can take a second tip from the field of journalism. Since late-breaking news may necessitate drastically cutting the length of an earlier written piece of news, journalists are taught to move main points up into the earlier sections of their stories and into the first sentence of each paragraph.

When you write to inform, think of what you are writing as a news or press release. Get your really important points out in your first paragraph, and then expand these points in subsequent paragraphs, each beginning with a topic sentence.

End good news messages on a positive, friendly note that completes the information you have just given your reader. If dates are important, you may want to repeat important dates. If your message contains special information, you may want briefly to remind your reader about that special information. If you want to build good will with your reader, do so in your closing.

When well written, your endings can be used for public relations, but only if all the important information has come first. The danger in giving your readers good news is that you will waste too much time getting to your point. Eliminate anything unnecessary from the opening paragraph, and save everything that is secondary until last.

Here's a simple message from an insurance company responding to several questions a policyholder had about an insurance policy. The questions and their answers are simple, but the message wastes time getting to the answers the reader wants. How would you rewrite it, using the PAFEO process?

```
                                        October 12, 1986


Frances Remo, M.D.
1252 Sixth Avenue
San Diego, CA 92101            RE: Policy # 879776543


Dear Dr. Remo:

Your letter of October 6, addressed to our President,
William DiPrimio, has been received. Mr. DiPrimio has
asked me to respond for him to your request for
information about your policy. We welcome the opportunity
to be of service to you.

Your policy has been paid in full as of September 21,
1986. There are no outstanding loans against the policy.
Our records list Mrs. Arlene Remo as sole beneficiary.

                              Sincerely,

                              Catherine Walters

                              Catherine Walters
                              Records
```

Organizing Bad News Messages: The Direct Approach

"A spoonful of sugar makes the medicine go down," according to Mary Poppins. "Lay it on the line," says the hard-nosed businessman. Which way do you prefer to get bad news?

Many psychologists believe it is better to lead the reader gently to the bad news, softening it first by adding a buffer and some explanation. But some writers prefer letting the reader have it right between the eyes. Whose spirits have not been lifted by those warm words, "We regret to inform you that . . ."? Is there ever a time when the direct approach is the best way to give the bad news?

You *can* organize bad news messages in the same way you organize good news messages, putting the most important information first, but only if customer goodwill is no longer important to you or if your next contact with your reader involves litigation.

In the following example, the writer has given a customer ample opportunity to pay a bill. All the writer's previous efforts have failed. The customer has not offered any compromise plan for payment. The writer no longer wants any future business from the customer, so the writer can send the following letter putting the bad news first. In fact, the letter is simply a statement of the bad news.

```
Dear Mrs. Wachman:

Your delinquent account, now 120 days past due, has been
turned over to a collection agency. Previous efforts to
resolve this matter have failed.

Sincerely,
```

Legal considerations may dictate the organizational pattern you use for your bad news messages. When the courts or arbitration will be used to resolve an issue, your statements of bad news should be direct. Your attempts at explanation could be used against you in any further legal action.

The direct approach to bad news has limited use. Most of the time, you will want and need to be more diplomatic. You will need an indirect or inductive approach to your bad news message because you will want to maintain as much goodwill as possible.

Organizing Bad News Messages: The Indirect Approach

Informative messages bringing bad news are especially difficult to write, since you run the risk of offending your readers. While you may for a variety of reasons wish to tell your readers some bad news, you will usually not want to lose your readers as clients or customers permanently. You must, therefore, try to soften the effects of the bad news you have to give by using psychology and tact. Your approach will consider the reader's possible reaction to your message.

In most cases, you will want to build up to, rather than start off with, your main point, the bad news. In good news messages, you begin with your strongest point, the good news, and you save considerations of reader rapport and public relations until last. In bad news messages, you begin with considerations of reader rapport and public relations because they are your strongest points in countering the bad news you will eventually deliver. The usual structure for a bad news message consists of these four parts:

1. A buffer statement.
2. The reasons explaining the disappointing message.
3. A clear and unmistakable statement of the bad news.
4. A goodwill message indicating a desire to maintain friendly relationships.

BUFFER   Buffers identify the general topic of your message without indicating that bad news is to follow. Here are three sample opening sentences that could be used to buffer bad news:

FROM A LETTER REJECTING A JOB APPLICANT:
Thank you for applying for a position in our accounting department.

FROM A NOTICE ANNOUNCING A RATE INCREASE:
We appreciate your patronage and are dedicated to providing you with the best possible service.

FROM A LETTER TELLING A CUSTOMER THAT SHE IS AT FAULT FOR THE MALFUNCTION OF A PRODUCT:
Thank you for writing to us about the problem with your car battery.

Each opening sentence identifies the subject of the message to follow and nods in the direction of the reader by expressing thanks or appreciation.

EXPLANATION   After the initial buffer, analyze and explain the circumstances that led up to the negative message you will give your reader. Such analysis or explanation must, however, reflect the reader's view of the issue at hand. Don't expect your reader to be impressed with the fact that your costs are increasing. If you can equate your increasing costs with a possible decrease in customer service, then you may get your reader's attention and understanding.

BAD NEWS STATEMENT   After the buffer and explanation, state the bad news in a way that cannot be misunderstood. Be firm, but reconsider your options one last time. Ask yourself if the situation must involve a "yes" or "no." If so, say "no," and be done with it. If the situation allows for a compromise by which both you and the reader can win a little, consider the compromise in the interests of goodwill. If you can't offer a compromise, try at least to offer the reader something positive as you also offer the negative. The following letter from a seed company illustrates how this can be done.

Spring, 1986

Dear Friend:

   Thank you for requesting a free copy of our seed and nursery catalog.

   Unfortunately, we've completely exhausted our supply of 6,000,000 catalogs--it seems every year, more and more

gardeners like you want to order their seeds from
America's most complete seed and nursery company, and this
year we've run short of catalogs. Your name has been added
to our list for future catalog mailings, though, and
you'll be sure to receive our big 1987 Spring Catalog.

Even though we can't send you our full-line catalog, we
want you to have the 10-page booklet of "Thank You"
Bargains enclosed. These special offers are normally
available only to customers who have already placed an
order, but since we can't comply with your catalog request
we want you to have a chance to take advantage of these
special bargains.

We're also enclosing a packet of Marigold seeds--at no
charge, of course. It's just our way of saying "Thank You"
for your interest in our products.

                                   Sincerely,

GOODWILL STATEMENT Take careful note of the last paragraph in the seed
company's letter. After reading it, don't you feel that these are pretty nice
people? The letter ends with a sentence that generates goodwill. You almost
forget that they said "no" to your request.

A goodwill statement protects the reader's self-esteem. The reader is better
able to accept bad news if there is no element of personal rejection or
rebuff in the letter. The reader's self-esteem is further protected when the
writer makes an effort to continue a business relationship with the reader.

Let's have a look at one more example of a well-planned bad news letter.
A college professor wrote asking for thirty copies of a company's annual
report to use in his business writing classes. Here is the reply he received.
Does the letter contain all of the elements of a bad news letter discussed
above? Can you identify each?

Dear Professor Hannum:

Thank you for your letter of April 27 requesting 30 copies
of our annual report for use in your classes.

We are flattered that you feel our report is so well
written and wish we could accommodate your request.
However, the number of such requests from colleges and
universities, both in this country and abroad, for
multiple copies to be used in accounting and finance
classes, has been so great that we simply cannot comply.
We have adopted a practice of supplying only one or two
reports and asking the institution to copy whatever
portions they need.

I am sorry that we can't fulfill your request but hope you can understand our reasons for not doing so.

                                                          Sincerely,

Instructions

One of the most important and most common forms of informative writing is the writing of instructions. These may range from the instructions on the label of a paint can to those outlining the procedure for assembling your own computer. Instructions may be one word (Stop!) or a whole book (for example, the manual accompanying a word-processing program). Their length is not, however, the challenge to the writer. The real challenge in writing instructions is to make them clear, concise, and complete, so that the reader cannot possibly misunderstand them. If you have ever tried to write directions to your home for a visitor from another state, you have some idea of how difficult writing instructions can be. You also know the important fact that faulty instructions produce immediate consequences: your guests never arrive.

Clear, concise, and complete instructions are necessary for the efficient operation of any business. Procedures must be written down so that every new employee does not have to reinvent the wheel. Instructions are important to the safety of employees and to the legality of the company's actions. People need instructions to assemble and maintain equipment and to operate machinery. Are you getting the message that instructions are one of the most basic and essential forms of writing? They are indeed, and in the following paragraphs, we'll show you how you can use the PAFEO process to help you write instructions that work.

Purpose

You write instructions to tell your reader how to do something or how something is supposed to work. The implication is that you know how and your audience doesn't. The reader wants to hear the voice of authority, the tone of confidence. Write instructions in the imperative mood—the grammatical term for a command, not a declaration or a question. (The preceding sentence is an example of the imperative mood.) Your operational purpose requires that you persuade the reader that the steps you propose will lead to success. The thought behind the instructions goes like this: "Trust me. I've done this before. If you do exactly what I tell you, you can't go wrong."

In keeping with your purpose, keep your sentences short, and use the active voice. Label transitions and steps clearly with signposts like *first, next,* and *then.*

Audience

When you write instructions, you are playing the role of the expert; the audience is largely uninitiated. That situation is a special kind of trap for the writer. Because you know what you're talking about, you must exert much effort to recover your reader's innocent state. In many instances, it seems all but impossible for the expert to shed all that expertise and get back to the point where he or she does not know how to turn the machine on. Yet that's exactly what must be done in a great many instances.

To write useful instructions, you must give careful thought to the amount of experience and sophistication your readers will bring to their reading. When you have a range of readers, some more experienced than others, it is best to aim at the least experienced, least sophisticated among your potential readers. Have you ever heard anyone complaining that instructions were too basic, too clear?

The manuals accompanying computers and software programs have generated much criticism from audiences who find them frustrating and often incomprehensible. Although these manuals have improved greatly in recent years, many still leave the reader unsatisfied and unhappy. Proof lies in the success of books written by experienced writers dealing with the most popular software programs. These books know how to reach the audience, most of whom are learning to use a word-processing program for the first time. The manuals, on the other hand, are written by programmers and computer science specialists who have difficulty remembering that the reader may be totally unfamiliar with even basic computer jargon. Page 2 of one manual tells the reader that step 1 is

1. Boot up your system with your DOS disk.

If you have some experience with computers, that instruction is clear. If you have just spent several thousand dollars to buy a computer and several hundred more on the word-processing software, you may want to boot them both, along with the manual and its author.

Format

You can present instructions in paragraphs or lists. The choice of the most appropriate is up to you, after due consideration of the material, your purpose, and your audience. Here is an example in paragraph form:

TO MAKE A CORRECTION
Separate the master sheet from the carbon sheet. Use a razor blade to "lift" the image deposit that forms the error. Avoid damaging the finish of the master. Then rub gently but firmly over the entire error with an eraser or correction pencil. An eraser shield may be used to isolate the error.

Bullets or numbers are usually used with a list. Use numbers to emphasize steps in a sequence, bullets to emphasize separate items. Here is an example of enumeration:

1. Take two sheets of paper large enough to cover the pages of the book when folded in half.
2. Measure the pages of the book, mark the paper, and cut off the excess.

Evidence

The information in your instructions must be both clear and complete. But how complete is complete? Instructions often must be presented in a limited space, and most readers have limited time to spend reading them. Nevertheless, you have to take care not to omit anything as too obvious—too obvious to you, that is—by returning to the analysis you have made of your audience. The evidence you need is the information your least sophisticated reader needs to complete the task successfully.

Organization

Instructions are generally organized in a step-by-step pattern. Be careful not to leave out any steps. Such an omission is the most frequent cause of consternation among readers. After you write the instructions, follow them yourself in carrying out the procedure. You will quickly discover whether you have forgotten any of the steps.

Longer instructions may employ an introduction that provides necessary background, explains the purpose of the procedure or the mechanism, and defines terms.

Depending on the nature of the instructions, you may also need a list of equipment or materials needed. Optional items should be identified in parentheses, as in this example:

To use this program on your personal computer, you need:
1. A Super 100 PC with at least 128K memory.
2. At least one disk drive.
3. An 80-column monitor.
4. An operating system disk.
5. The Priceless Word distribution disk.
6. Several formatted, blank disks.
7. A mouse (optional).
8. A printer (optional).

Adjustment Letters

Purpose and audience are especially important considerations in writing adjustment letters responding to customers' claims or complaints. The purpose of the letter is to retain customer goodwill by granting any reasonable adjustment. Since the purpose is to create goodwill, the tone in which the adjustment is made carries much importance. Your letter is a good news letter. Your audience must not feel as though you are grudgingly giving them something they don't deserve; such a tone will undo all the good of the adjustment. The messages you want to convey are:

- We care about you.
- We're sorry there was a problem.
- Here's what we are going to do about it.
- Here's our explanation of what went wrong.
- Thanks for letting us know.

Don't, however, try to follow a formula; make the letter personal and warm. You should have no trouble writing adjustment letters if you remember that your real purpose is to build goodwill, if you keep your audience's perspective in mind, and if you organize as you would any good news letter. The following letter is a positive example of an adjustment letter answering a complaint from a dissatisfied candy eater.

Dear Mr. Piccoli:

Thank you for your interest and the time taken to relate your recent experience with an Alice candy bar.

Alice candy bars, like many other fine food products, are best when they are the freshest. Based on the stale condition of the Alice bar you described in your letter, we would surmise that in all likelihood the bar was on the retailer's store shelf for too long a period of time.

To correct this situation we ask that you provide us with the store name and location from which you purchased this product, so that our representative can investigate.

Please accept our apologies for any inconvenience this may have caused you. As an expression of our goodwill, a sample package of candy bars will be forwarded to you under separate cover.

Yours very truly,

The "Memo to the File"

On the job, situations may occur that require some kind of written record for future reference. One example would be a personnel appraisal. The easiest way you can insure such a record is to write or dictate a memo to the file. In writing such a memo, you should follow the guidelines for direct informative writing discussed earlier in this chapter. Organize your information deductively, and pay careful attention to the problem of audience. The reader may be consulting this memo years later and may have no direct acquaintance with the person or situation.

Generally, memos are put in the file so that they will be read by someone other than yourself. Don't forget to anticipate any problems your readers might have with your memo. Your readers may not have been present when the incident you are writing about occurred. Even if your readers were present, they may need to be reminded about the facts of the incident. Be specific then and explain who, what, when, where, why, and how with details that will make sense to your intended readers.

Here's a memo intended for an employee's personnel file:

```
TO:     File                      DATE: February 21, 1987
FROM:  K. L. Stein,
        Supervisor-Customer Service
RE:     Outstanding Performance by
        Stephanie Wood

I wish to commend Stephanie Wood for outstanding work
performance.

The blizzard on February 12 and the three-day clean-up
that followed made it impossible for many employees to
make it into work.

Not only was Stephanie on time for these four days but she
also agreed to work an extra four hours each day. During
her overtime, she also assumed supervisory responsibility
in addition to her regular duties as a service
representative.

That the department continued to function smoothly during
this difficult period is due in large part to Stephanie's
efforts.
```

Messages Introducing New Services or Products

Whenever you write to inform about new services or products, you have to be especially careful to anticipate any difficulties your readers may have. Remember that you are the expert. You already know about the new service or product. Your audience is, however, in the dark. Your primary purpose may be to persuade the reader to use the service or to buy the product, but you will have an easier time persuading if you inform first.

Here are several points to keep in mind whenever you write to introduce a new product or service:

1. Start at the beginning and explain *everything*.
2. Use a step-by-step approach in your explanation that anticipates and then solves any problems your reader is likely to have.
3. Consider using visuals to inform your reader. A picture, chart, or drawing can be worth a thousand words.
4. Tell your reader everything he or she needs to know, especially where to go for more information or help.

The following illustration accompanied all customers' monthly bank statements as an enclosure.

Recently some banks have been doing a lot of advertising about high interest rates and big tax savings. Unfortunately, we believe most of the ads have been confusing and in some instances perhaps a little misleading. For this reason, we feel it is our responsibility to explain the facts about this unusual situation so you can make the investment decision that is right for you.

The New Law

Congress recently approved a new tax-free interest savings certificate with the following features:

- Starting October 1, 1981 and for the next 15 months, all banks can offer this new certificate.
- Some of the names used to refer to this new certificate are "All Savers Certificate" or "Tax-Exempt Certificate." We refer to ours as the All Savers Certificate.
- The term of the certificate is one year.
- The interest rate should be lower than the rate paid on 6-Month Certificates or the 2½-Year Investment Certificates. The reason why the interest rate may be lower is because of the unique tax-exempt feature.
- Each individual is allowed a one-time exemption of $1,000 of interest, $2,000 for people who file joint tax returns.
- A new interest rate, which is 70% of the prior month's one-year U.S. Treasury Bill investment yield, is announced every four weeks on a Thursday. The rate becomes effective on the following Monday.
- Many banks will require a $500 deposit to purchase this new certificate even though the law does not require it. However, EAB's No-Minimum-Deposit All Savers Certificate can be purchased for any amount. This way, everyone can take advantage of this new tax benefit.
- All banks will be allowed to offer the same interest rate. No bank, including savings banks, will be able to pay higher interest than EAB.

What This Means To You

First, please do not invest or sign up for any investment program unless the institution you are doing business with is willing to give you the terms and conditions of your investment in writing.

Find out all your investment options. You should be aware that the All Savers Certificate is not for everyone. The higher your tax bracket, the greater the tax benefit you can derive from the new certificate. If you overinvest and earn more than the $1,000 or $2,000 lifetime exemption allowed by law, you might reduce your maximum earnings.

We believe the All Savers Certificate is one of the best ways to keep more of the money your investments earn. However, I cannot stress the importance of knowing all your options in order to invest wisely. To insure that you are earning the maximum income possible, I encourage you to stop by any of our branches to personally select a total savings plan that best meets your needs.

If you have any questions about EAB's All Savers Certificate, our 6-Month Certificate or even our 2½-Year Investment Certificate, just call our Banking Center from 9 A.M. to 9 P.M., Monday through Friday:

(516) 248-7020   (212) 895-3404   (914) 761-6400

We will be glad to answer any of your questions and if you would like, we will send you an application so you can purchase your All Savers Certificate or any other EAB Savings Certificate conveniently through the mail.

At European American Bank we work hard to offer you the finest in banking products and the very best in personal service.

Reprinted by permission of European American Bank.

This message from the bank's president announced a new product, the All Savers Certificate, that the bank was offering. While providing a good deal of information to the customer, the enclosure may also persuade the customer to buy such a certificate. (We will have more to say about persuasive messages in the next chapter.) Note too how this enclosure is an excellent exercise in public relations for the bank, the European American Bank (EAB).

Exercises

1. Using the PAFEO process, discuss the effectiveness of the following informative messages. Rewrite any your instructor directs you to rewrite.

SAMPLE A

MEMORANDUM

Department of Health, Education, and Welfare Social Security Administration

TO: All Managers   DATE: June 16, 1981
FROM: Director of Management   REFER TO: SPRF32
SUBJECT: Sexual Harassment Training

The Sexual Harassment Training scheduled for all employees has been postponed until further notice.

SAMPLE B

April 24, 1986

TO: All Faculty
FROM: Joanne Butler, Dean
SUBJECT: Final Examination Schedules

Final Examinations in the College will be held during official exam week, May 5 to 9. The examination schedule is printed on page twelve of the Spring 1986 Directory of Classes.

Please abide by that schedule.

Resist all temptation to give exams early. Stand firm in opposition to your students' pleas; they are misguided.

Help your colleagues to remain firm in their resolve to give exams only at the scheduled times; some may be weak and need your support.

If there are compelling reasons for giving exams at times other than those scheduled, talk it over with your chairman and get specific permission to make the change.

Chairmen: Send on to me the name, and the course number, of anyone to whom you give permission to give an exam outside the scheduled time.

SAMPLE C

Dear Subscriber:

We hope you are enjoying your copies of SATELLITE Magazine and that you're finding them enjoyable, interesting and informative.
And now, we'd like to ask you to do us a favor.

    Our records show that we have not yet received payment for your subscription.

Won't you please take a moment now to send us your check for the enclosed subscription bill. An envelope is provided for your convenience.

You'll save us additional expense, and you'll be sure of receiving your copies of SATELLITE Magazine without interruption.

Please send us your payment today.

Cordially,

Karen Johnston
Circulation Manager

P.S. If you recently mailed your payment, it has probably crossed in the mail with this note. Please accept our thanks and be assured your account will be properly credited.

KJ/bnc

SAMPLE D

TO OUR READERS
Effective today, the price of the Sunday Bulletin is $1. The price of the weekday Bulletin remains 25 cents. In raising its Sunday price to $1, The Bulletin joins other leading metropolitan papers.

The cost of just the paper required to print a typical Sunday edition is 77.5 cents. There are other substantial costs for editorial content, printing and distribution. Thus, the cost of publishing each Sunday Bulletin is almost double its new $1 price. Advertising revenue, of course, offsets some of this expense.

Part of the price increase goes to the carriers, dealers and others who distribute The Bulletin. Their costs also have been rising.

The management of The Bulletin continues to honor its commitment to improve the value of this newspaper to its readers. In the last year, the editorial staff has been strengthened by nearly a hundred reporters, editors, photographers and support personnel. News coverage has been expanded in the metropolitan area, the nation and the world; The Bulletin's network of bureaus now reaches Moscow. At the same time, The Bulletin has improved its specialized coverage from sports to food, from business to entertainment.

Over the last two Sundays, a bright new format and a new name ("Bulletin") also have been introduced for the magazine prepared exclusively for readers of the Sunday Bulletin, and TV magazine has been improved strikingly.

SAMPLE E

April 7, 1986

Dear Candidate:

I am sorry to inform you that the Search Committee for the Vice President/Academic Dean's position at the College is in the process of selecting a final candidate. We had many well-qualified candidates and the choice was difficult.

Thank you for your interest in the College.

Very truly yours,

SAMPLE F

May 21, 1986

Mrs. Frances Parola
398 Lighthouse Road
Wilmington, DE 19609

Dear Mrs. Parola:

Thank you very much for making us aware of the problem you found with our oatmeal cookies. We are concerned about the quality of our products, and so we appreciate having this brought to our attention.

If you have used our products before, I am sure you know how hard we strive to see that our customers receive cookies, crackers, and snacks which are manufactured according to strictest standards of control. We try not only to meet, but to exceed all standards set by federal, state, and local regulatory agencies. Our finished products have to pass through many quality control checks before they can leave our bakeries. Because every effort is made to constantly strive for quality, purchases like yours are a concern to us.

We have analyzed the sample you sent to us and have identified it as a harmless accumulation of ingredients that did not become part of the dough. Although this is certainly undesirable we are glad to relieve you of any concern you may have had. The bakery that manufactured this product has been notified of your dissatisfaction. They are investigating and taking steps to ensure this will not be repeated. A description has also been put into our weekly report which is thoroughly reviewed and acted on by our corporate and bakery levels. We assure you this matter will not be taken lightly, and we will do everything we can to make certain it does not recur.

Please accept our sincere apologies for your disappointment in our product. I realize the enclosed coupon amount cannot make up for the unpleasantness of your purchase, but we ask that you accept it as our way of thanking you for being an interested customer, and for being concerned about the quality of our products.

Sincerely,

SAMPLE G

OFFICE MEMORANDUM

TO: DATE:
FROM:
SUBJECT: Out Patient Services with Paducah Health Plan
          (PHP) Insurance

Enclosed you will find a copy of your bill that has been changed to self pay. This means that it is your reposnsibility for payment of this bill. We have not received payment from PHP and are no longer accepting this insurance. You may submit payment to us directly and then submit the claim to PHP for possible reimbursement. If payment for these services are not received within a certain period of time, your account will be scheduled for a collection agency. Your attention to this matter would be greatly appreciated.

SAMPLE H

OFFICE MEMORANDUM

TO: Distribution
FROM: L. Janet McClure-Davis
RE: Recent Systems Errors

On June 10, 1986 a systems error was made on Withdrawals of all Discretionary Common Trust Funds. An entry purchasing back the Units was posted a few days later, and the units were then withdrawn correctly. For Court Accounting and Distribution purposes the Admission buying back the entry should be treated as a Cancellation of the original June 10, 1986 withdrawal, and only the correct entry should be stated by the Accountant. A similar error was made on the admissions of Discretionary Common Trust Funds on June 10, 1986.

SAMPLE I

OFFICE MEMORANDUM

TO: Bookkeeping
FROM: L. Hollander LaRue III
        Legal

The last several entries in an account represent our fee, counsel fee, filing costs, etc. which, although they are indicated as having been paid on a certain date, in fact were not paid on that date and are not paid until a subsequent date. It is my understanding that it is the practice of numerous banks to include these items but they do not put a date of payment in front of the entry. They have no difficulty with the courts concerning these entries. I see no objection to this procedure and it would probably be advisable in that it avoids the question of counsel that it appears he was paid on a certain date when he was not. Therefore, in the future please eliminate the date from the entry.

SAMPLE J

<u>M E M O R A N D U M</u>

TO: ALL INSTRUCTORS
FROM: Jane Dean
DATE: JUNE 1, 1986

It was very distressing to learn that almost all of the
students in the afternoon session chose to cut classes
last Friday. It was clear to everyone Friday morning, that
it was impossible to hold a picnic. Students should have
been cautioned that if the picnic was rained out, classes
<u>would</u> of course be held. I can not conceive of any
employees working for a regular company pulling this kind
of nonsense. Your students will soon be employees and they
must be made to understand their responsibilities.
Students out will be marked absent and will have to make
up the time missed. Moreover, it was even more
discouraging to learn that a number of instructors were
unprepared to teach class; this is a most serious matter
and warrants appropriate notation in individual teacher
files.

Weather permitting, we will have our school picnic this
Friday, June 4th. The same criteria will apply, in that
classes beginning 11 A.M., will be excused. Students are to
be told in no uncertain way, that if there is even the
slightest possibility of being rained out, they are to
call the school by 10 A.M. for confirmation. They are to
make no plans for Friday, other than to go to the picnic
or come to school.

If there are any questions about this matter, I will be
happy to speak with anyone personally.

SAMPLE K

January 16, 1986

Mrs. Marie Montclair
6074 Vineland Avenue
North Hollywood, CA 91606

Dear Mrs. Montclair:

We were surprised to receive your letter concerning our
Zesty Italian Salad Dressing. As one of our customers, you

know we are proud of the reputation for quality of all our products and take extensive care in our manufacturing processes to maintain this quality.

We are at a loss to explain the reason why part of the jar rim was missing from your jar of Zesty Italian Salad Dressing. Needless to say, we are concerned with your report and will have our Production and Quality Assurance personnel investigate this occurrence.

Enclosed is $1.50 to cover the cost of your jar of dressing. Also enclosed is an assortment of recipes for your enjoyment. Under separate cover, we are sending you an assortment of our other products which we hope you will enjoy trying.

Thank you for bringing this matter to our attention. We are sorry for any inconvenience caused you.

Cordially,

2. Write an explanation of the requirements for graduation for students majoring in your discipline. Your audience is made up of high school seniors who will attend your school next year but who are undecided about a major.

   (Variation on this assignment for adult students who work full-time and attend school part-time: Write a letter to a college freshman who is interested in knowing what courses to take to prepare him or her for an entry-level position with your company.)

3. Prepare an announcement for a new product or service or for a special event on or off campus.

4. You have been asked by the president of your school to head up a campus drive to raise funds for a worthy cause. Write a letter to the president declining the request.

CHAPTER 8

Messages That Persuade

In the previous chapter, we discussed at length messages that inform. Some of these messages were purely informative; others combined information with a secondary purpose of persuading or building goodwill. In this chapter, we'll discuss messages whose primary purposes are to persuade, although any business message can have an element of persuasion.

Typical of such persuasive messages are these:

- Sales letters.
- Requests for an adjustment or a favor.
- Internal proposals aimed at instigating a change in policy or procedure.
- External proposals (sales or grant proposals).

What each of these forms has in common is a desire to do more than simply inform; each attempts to move an audience to act in the way the writer wants. Social scientists and marketing experts have devoted much time and effort to analyzing just how people are persuaded. They have devised approaches that appeal to audience's egos, emotions, or wallets. In doing so, they have refined the idea of purpose into a sales formula that can be useful to you whenever you have to write an unsolicited persuasive message. It can be useful, that is, so long as you remember that it is a guide to the purposes of a persuasive message, not an absolute formula to be followed slavishly and unimaginatively.

The AIDA Formula for Persuasive Messages

By this time, you are familiar with the PAFEO mnemonic. At the risk of becoming an advertisement for alphabet soup, we'd like to have you look at purpose in terms of a more widely used mnemonic, AIDA, which will help you remember the four steps in accomplishing your persuasive purpose. Whether your purpose is to sell a product, a service, or an idea, the AIDA formula can serve as an outline for constructing a persuasive presentation. Here's the formula:

A   Attention
I   Interest
D   Desire
A   Action

AIDA can be used either for hard sell (in which writers make their pitch first) or for soft sell (in which they build up to their pitch after informing their audiences).

Advertising makes use of both approaches, often with mixed results. Persuasive writing can also use either approach, depending on audience, format, evidence, and organization, the remaining parts of the PAFEO process. But hard or soft sell, your persuasive message must begin to persuade right from the start. Failure to do so will ensure the failure of the overall persuasive message.

As an example of how not to start a persuasive message, consider the following opening paragraphs from a proposal from an English department. The department requested a grant of $5 million from a foundation that had solicited a proposal for a program to raise the level of literacy in the region surrounding the university.

If you are confused about the writer's sense of purpose and audience, you're not alone. The foundation denied the proposal, even though it had originally sought such a proposal from the university. Without using what follows as an excuse for revenge on your past writing teachers for wrongs imagined or real, critique these paragraphs using the PAFEO process and then the AIDA formula to evaluate its selling power. Would you say these paragraphs are reader-based or writer-based? Why?

> The lack of skill in reading and writing among young Americans has now caught the notice of the nation. The situation is indeed appalling: hardly a college, no matter how selective, can today do without a remedial English program. Even college mathematics instructors suspect that their students cannot solve elementary problems because they cannot read them accurately. Everywhere demands are heard that something be done, but the reactions to date have been

predictable: commentators select scapegoats—school teachers, parents, television programming, or the discipline of English itself; various agencies appoint committees and commissions; school officials demand more homework. But despite the loud alarms, there has been no massive national response. When the Russian success in space alarmed our people over the state of American science, the response was quick and urgent because it was clear to the scientific community what needed to be done. Faced with a threat to literacy itself, however, humanists today differ not only over goals and means, but even over whether anything can be done at all. We wait immobilized as the skills of literacy bleed away. The appointment of a Dean of Writing or the imposition of more degree requirements is of little more than symbolic value. Nor is nostalgia much help. Defining the educated person as one who has completed Cicero sharpens the definition and not the student.

The collapse of those skills basic to learning did not occur by itself. Something has happened in our society of which this collapse is but a symptom. Changes in the family, the aftereffects of Viet Nam, the pervasiveness of television, all have affected the attitudes of the young and their assimilation of their culture. These changes have largely taken place during the past decade. During that period America also steadily broadened the base of its democracy. Disadvantages of race or sex or class, of circumstance or birth, of exploitation or accident or fate— these the nation set itself to put right, not only to acknowledge the legitimacy of the hitherto unvalued, but to bring them into the "mainstream" of our society. This renewed insistence on the radical import of the American dream coincided with a widening discontinuity in our culture. Thus, at the very time we broadened access to education we lost a definition of what it was to be an educated person. We lost the essential connection between the American community and the American academy.

We see today the result in student helplessness. A decade ago we welcomed minority students from the inner city, only belatedly realizing that in initiating them into the language and lore of the liberal arts we might be asking them to repudiate the culture which supported their identity. A few years later we welcomed the so-called "new student," and then realized that this child of blue-collar parents saw elements of a high culture as but obstacles to the job training which paid in dollars. Today, we find even the traditional student often rootless. Quiet campuses and businesslike demeanors only mask confusion and diminished expectations. Students read perhaps half as much as they did a decade ago, and their instructors, while deploring their students' lack of writing skills, are forced by the sheer labor of deciphering student essays to curtail still further their assignment.

The tragedy is both societal and personal. The skills of reading and writing, at the level endorsed by the humanities in colleges and universities of good quality, are the educational keys to personal effectiveness and power. Access to wider cultural choice, one supposed purpose of education in America, depends upon a person's control of his language. We all realize that in the act of writing every writer joins a stylistic community. "Better" writers (and speakers) join a "better" one. Thus the student who cannot use his language is denied both access to power in his society and cultivated choice. Within a broadened democracy, that is the real pathos of our inability to teach matters of "style" to students intent on getting ahead. So disabled, we teach them almost nothing at all.

Crash efforts to retrain teachers and administrators only serve to publicize the extent and urgency of the problem rather than arm us to solve it. For we have no received method of teaching English. Despite the journals devoted to language and composition, the field of thought remains unfocused and undisciplined. Writing instructors in this country are almost always trained in literary criticism, but in practice the applicability of such training is inhibited or denied. Most rhetorics and handbooks display a remarkably thin and impoverished notion of the aims and effects of writing, and the teaching of composition remains an *ad hoc* art dependent on certain unquestioned assumptions of sanction and taste. Given the loss of those shared certainties, given the lack of any science of remediation, we face a crisis not unlike that which confronted traditional medicine in the fourteenth century, an *ad hoc* art dependent on outworn authorities in trying to cope with the plague.

A plague is upon us.

A plague, indeed! Had enough? The people at the foundation did. They rejected the application, and they didn't encourage the university to reapply.

Attention

In order to persuade a reader, you must first get that reader's attention. Here are five time-honored ways of attracting your reader's attention:

1. Solve a problem for the reader: *Here's how you can save 20 percent on your computer costs.*
2. Ask the reader a question: *Are you having problems writing the papers you need to write for your courses here at the university?*
3. Refer the reader to a present event: *Right now, the committee is meeting to determine the salary structure for the coming contract.*
4. Use the reader's name, but only once: *Studies show, Mr. Thornton, that careful retirement planning really pays off.*
5. Alert the reader to an alarming fact or statistic: *As of April, 1986, there were 19,800 reported cases of AIDS. Of those cases, 54 percent of those diagnosed have died. No one has recovered from AIDS. Even conservative estimates indicate that another 250,000 people have already contracted the disease, many without knowing it.*

Interest

Once you've attracted your reader's attention, you need to develop that reader's interest. The easiest way of developing such interest is to describe your idea or product and to present evidence in support of what you propose. Keep these three approaches in mind:

1. Involve the reader from the start: *We're happy to report that you can have a new library built to your specifications for 20 percent less than our original estimate.*
2. Explain benefits in terms of your reader's needs: *A word processor will allow you to edit your document without time-wasting recopying by hand.*

3. Be specific about how and why the product or service will be useful, money-saving, or convenient: *This course is designed to show you how to write clearer and more effective letters and memos.*

Desire

To create desire in your reader, you need to make your reader want whatever you are offering. You can make your case more convincing by doing the following:

- Looking carefully at where you place your strongest points (whether they come first or last will depend upon your audience's expected reaction to what you propose).
- Including testimonials from people your reader respects.
- Restating your main points when necessary, especially in longer documents.
- Varying your pace so as not to overwhelm your reader and thereby discourage rather than encourage the reader to adopt your proposal.

Action

If you have managed to retain your reader's interest thus far in your persuasive message, you need to tell your reader how to act upon your proposal. In giving your reader these instructions,

1. Make sure you are specific: *Call me directly at 897-5555, ext. 345, if you need any additional information.*
2. Make the action simple: *To place your order, simply dial 1-800-999-8888; operators are standing by to take your order.*
3. Encourage your reader to act immediately: *This offer is good for the next two weeks only.*

The illustration on page 178 is an example of a sales pitch that uses AIDA rather effectively. Note also how the postscript is used as another way of prodding the reader to act.

Audience Analysis and Persuasive Messages

Audience analysis is essential as part of the necessary preparation for writing a persuasive message. Two convenient rules of thumb to remember are the following:

1. When your audience will be enthusiastic or neutral to what you propose, state your case in full up front.
2. When your audience will be hostile to or have problems with what you propose, build your case carefully, saving your full case until last when you have had a chance to win over the audience.

The Pentagon Papers contain an example of the latter approach. In 1961, General Maxwell Taylor sent President Kennedy a cablegram urging a course

Dear Valued Customer:

How would you like to receive $50 for emergency expenses if your car is in an accident? How would you like to save on just about everything you purchase for your home or apartment?

Now you can get all these savings and more when you join the *Bankard Club*, a division of JTX Club, Inc. Membership in this travel club and discount plan will bring you big savings and services covering many aspects of your life – from travel and recreation to everyday activities.

Discount services include hotel and car rental bargains... brand-name merchandise at big savings... an auto buying service... and savings on prescription drugs and vitamins.

Travel benefits include a Custom Travel Service... auto trip planning... travel insurance... road emergency and legal defense reimbursements... lost key and luggage return services. Plus you'll receive a new magazine devoted to travel and the leisure life.

All these benefits – and many more – will be yours for just pennies a day. Your dues will be conveniently billed each month on your Visa or MasterCard account, subject to normal credit standards. If you have any questions, call Bankard Club: 800-523-1965. (In Pennsylvania, call collect: 215-241-3223.)

It's easy to take advantage of this special invitation. Read this fact-filled brochure, and return the attached Acceptance Certificate. You'll soon start saving!

Sincerely,

Tom Joyce
President, Bankard Club

P.S. Please reply by April 30, 1981 and receive an Identifax kit as a free gift. Identifax is a system for coding your personal property. It also includes stickers to alert would-be thieves that you're protected.

Reprinted by permission of the Bankard Club, a division of JTX Club, Inc., and Mr. Thomas Joyce.

of action in South Vietnam (SVN). Anticipating some opposition from the President or his other advisers, Taylor began his message specifically but cautiously:

> This message is for the purpose of presenting my reasons for recommending the introduction of a U. S. military force into SVN. I have reached the conclusion that this is an essential action if we are to reverse the present downward trend of events. . . .

Later in the cablegram, after making his case and repeating his main point that the United States should commit forces to South Vietnam, Taylor delivered the most difficult point in his message:

> By the foregoing line of reasoning, I have reached the conclusion that the introduction of [word illegible] military Task Force without delays offers definitely more advantage than it creates risks and difficulties. In fact, I do not believe that our program to save SVN will succeed without it. If the concept is approved, the exact size and composition of the force should be determined by Sec Def [the Secretary of Defense] in consultation with JCS [the Joint Chiefs of Staff], the Chief MAAG [the Military Assistance Advisory Group] and CINCPAC [the Commander in Chief, Pacific]. My own feeling is that the initial size should not exceed about 8,000.[1]

Given Taylor's position and his opening paragraph, neither the President nor any of his other advisers is going to mistake him for a dove. But for Taylor to have led with a sentence such as "We should send 8,000 troops into SVN immediately" would have been "suicide by cablegram," since he needs first to make a case for so large a commitment of troops to this area.

Again, carefully assess the reaction your reader is likely to have to what you propose and proceed with caution. Should you lay all your cards on the table, or simply indicate which suit is trump?

Eleven Rules of Persuasion You Need to Know

In his business writing book that we have cited earlier, David Ewing of the Harvard Business School lists eleven rules every persuader should know. Ewing's own discussion is based in large part on the early work of two social scientists, Marvin Karlins and Herbert I. Abelson.[2] These rules help you better prepare your reader for the persuasive message you are about to deliver.

Here are the eleven rules:

1. Consider whether your views will make problems for your readers.
2. Don't offer new ideas, directives, or recommendations for change until your readers are prepared for them.

---

[1] *The Pentagon Papers, As Published by the New York Times* (New York: Bantam Books, 1971), pp. 141–143.
[2] See David Ewing, *Writing for Results,* 2nd ed. (New York: Wiley, 1979), pp. 65–80; and Marvin Karlins and Herbert I. Abelson, *Persuasion* (New York: Springer, 1970), *passim.*

3. Your credibility with readers affects your strategy.
4. If your audience disagrees with your ideas or is uncertain about them, present both sides of the argument.
5. Win respect by making your opinion or recommendation clear.
6. Put your strongest points last if the audience is very interested in the argument, first if it is not so interested.
7. Don't count on changing attitudes by offering information alone.
8. "Testimonials" are most likely to be persuasive if drawn from people with whom readers associate.
9. Be wary of using extreme or "sensational" claims and facts.
10. Tailor your presentation to the reasons for readers' attitudes, if you know them.
11. Never mention other people without considering their possible effect on the reader.

Format and Persuasive Messages

As with any message, format is important for persuasive messages. Shorter sentences and paragraphs and ample use of white space help the reader digest the information you are passing along and set the reader up to do what you want the reader to do.

Certain kinds of persuasive messages—especially proposals to external sources of funding—must conform to rigorous formats before they will even be considered by the funding source. If you have to write such a proposal, check several times with the funding source about style and format and closely follow the guidelines you are given.

Evidence and Persuasive Messages

Do your persuading with specifics. If you're describing the appeal of the product, show the reader the special features that distinguish it from the ordinary. Will it make life easier? Be vivid in describing how. Will it save money? Show how it will help pay for itself in savings. Give amounts in the ways in which they sound most appealing. Costs are stated in "pennies a day," while savings are accumulated at the rate of "hundreds per year." But remember the wisdom of expert persuaders and be wary of using extreme or sensational claims or facts.

Testimonials can be a useful form of evidence to use in persuasion, but only if drawn from people with whom readers can associate. If you are selling insurance to the elderly, for example, you don't get anywhere by quoting testimonials from a rock star.

Sound figures, clearly presented, remain the most effective kind of persuasive evidence in business writing. Check and double-check the accuracy of any figures or statistics you present as evidence. When it is necessary to mention money in a persuasive message, be guided by the rules of thumb we mentioned earlier in the chapter. State the price up front if it constitutes a selling point, something the audience will be enthusiastic to learn. Otherwise, hold the discussion of costs until you have had a chance to build a strong case for the benefits and desirability of the product or service.

Organization and Persuasive Messages

General Taylor followed his initial statement of his purpose in his cablegram to President Kennedy with a list of *disadvantages* to his proposal. He also listed his disadvantages from least to most disadvantageous. Such an approach argues for an objective proposal. Readers are wary of your "bright ideas" and want to know you have considered alternative courses of action or both sides of the issue at hand.

If readers remember most of what they read last, it might not always be wise to lead with advantages and follow with disadvantages. But there are no concrete rules here; simply organize your message in a way best designed to accomplish your purpose.

Requests for Adjustments

When something goes wrong—an order that does not arrive, a product that is defective, or a service poorly rendered—it may be necessary to persuade your audience to make the adjustment you desire. Since your purpose is to produce an action that your audience is not anxious to take, your persuasive request will generally follow the indirect method you used in writing a bad news message.

It may, for example, begin with a general and neutral statement, a buffer that either party could accept (e.g., "Having been a long-time customer of the OneYear Tire Company, I know you take pride in the dependability of your tires").

The body of the letter will be your evidence supporting your claim. Be sure to include all relevant specifics such as date and place of purchase, purchase order number, and dates of previous correspondence.

End your letter with a specific request for adjustment. If you will accept several alternatives, list them. Keep your tone reasonable throughout the entire letter.

Here is an example of an adjustment request received by a paint company. The writer is a small businessman who does custom cabinet work in kitchen remodeling. What are its strengths and weaknesses? If you were Mr. Tyler, would you be inclined to make an offer of adjustment? Why or why not?

Dear Mr. Tyler:

After doing business with your company as a professional businessman for many years, I now find myself in a situation which we don't seem to be able to resolve. Because I use your paint products exclusively, I do not wish to jeopardize the excellent results your products give me, but I am at the point where I must insist on some resolution.

It began three years ago, when I took two of my Binks spray guns, #18 and #62, into your Springfield store for repairs. They were turned over to Jim Barrett. Subsequently I'd drop by to see if they were ready, and finally they were, but on that visit I did not have sufficient funds to get them out. Because of business pressure, I admit I did just use the other guns I had, and those two slipped my mind. The company did not send me a reminder or notice of intent to sell them. When I did remember last Monday and went in, I found that they had been sold. All the records had my name and phone number, and I am also listed in the phone book.

These guns were purchased new from your company a period before being brought in for repair. I would like to know just what your company would consider a fair settlement of this situation. I admit my own carelessness in this situation, but your company's failure to notify me makes it mutually at fault.

My attorney, Mr. Stephen Koch, can be reached at 212-4759. Please advise him of your feelings on this matter.

Sincerely,

Steve Nolen

Steve Nolen

Your instructor may want you to rewrite the letter, incorporating some of the suggestions developed in the class discussion.

Internal Proposals

Internal proposals are generally made in memo form, since they are documents meant to persuade someone at a higher level within your company. They are, therefore, less formal than a sales proposal directed at a potential client, but their purpose is nevertheless *to sell*. In an internal proposal, you are selling your idea to someone in a position to make it happen. The AIDA formula you learned to use in sales messages can be useful here too.

Careful audience analysis may mean the difference between success and failure in selling your idea. Think about the people who have to be convinced.

Which ones will resist your idea and why? Ideas, even good ones, are not always welcomed with open arms. New policies and procedures can be threatening. At the start, they usually entail extra work for people in authority, and untried ways contain an element of risk. Most of us prefer the security of doing things in the way we know rather than learning a new way. If you want to persuade readers to accept your new idea, you also have to persuade them that it doesn't threaten their own present security or status. Consider the psychological and political realities of your audience as you build your case. Ideas do not sell on merit alone. Your audience must be convinced that your idea has specific benefits, not hidden threats.

Typically, an internal proposal may recommend a change in procedure, the adoption of a new procedure, or an expenditure for needed equipment. The problem-solution organizational pattern can be used in combination with the AIDA formula. Here is how you might organize your proposal, using the basic plan of an introduction, body, and conclusion.

Introduction

Gain the reader's attention and interest by identifying a clear problem that needs solving. If the boss sees no problem, he or she obviously sees no need for the proposal, especially if it's a proposal that involves spending money. Your description of the problem must be detailed enough to persuade the reader that you thoroughly understand the problem and that you also share the reader's goals of cost efficiency, improved production, employee morale, company image, or whatever else your audience analysis reveals the reader is most concerned with.

When you have presented the problem, summarize your solution to it and the benefits you expect to be realized.

Body

The body of your proposal is a detailed, practical solution to the problem presented. How detailed? That depends on your audience's needs. You want your reader to make a decision. What kind of information would be needed to make that decision? Use the invention techniques you have learned in Part I to help you develop the information. In all likelihood, the reader will want a breakdown of costs; any requirements for personnel, material, and space; and a schedule for carrying out the proposal.

Conclusion

Keep your conclusion brief. You may want to summarize your recommendation for greater emphasis. End on a cooperative note ("Please let me know what you think"; "I'll be happy to answer any questions you may have"; "I'd be pleased to discuss this with you at your convenience").

Here is an example of an internal proposal. Does the proposal clearly present the problem? Is the solution sufficiently detailed and convincing? Does the proposal sufficiently anticipate possible objections from the reader?

TO: Joseph Baker, Vice-President
FROM: Joan Campbell, Office Manager
DATE: November 19, 1987
SUBJECT: Proposed Flex-time for Office Staff

Chronic lateness among the clerical staff is becoming a severe problem in this office. The payroll clerk reports that 20 of 25 secretaries and clerks have been more than fifteen minutes late at least ten days during the past month. Resulting problems include: resentment among on-time staff; inefficient processing of office correspondence and reports; and increased overtime pay for payroll clerk, who must spend an increasing amount of time fine-tuning timecards for late staff, in addition to her other duties.

Most of the latecomers are the new employees we hired last year when the company expanded; all have been well-trained, and are worth keeping on staff. Most, however, are mothers with school-age children. The current 8 A.M. clock-in time is too early for them to be able to get their children off to school <u>and</u> get themselves to work on time. I propose that we institute Flex-time for the clerical staff in order to solve this problem.

<u>What is flex-time?</u> Flex-time allows the employee to set the limits of his or her work day within time periods set by us. The employee may choose to come in any time between 7 A.M. and 9 A.M., and leave 8 1/2 hours later (allowing 1/2-hour for lunch), between 3:30 P.M. and 5:30 P.M.

<u>What are the advantages of Flex-time?</u> Flex-time would allow us to get eight hours of work for eight hours of pay from the clerical staff. It would not cost the payroll clerk any more time than the present system already does. It could be implemented quickly (by January 1, 1988). And it would improve relations between the employees and the supervising staff and among the employees themselves.

<u>Won't Flex-time promote all kinds of varying hours?</u> No. The idea is that each employee sets the limits of his or her average working day. The employee will determine his or her usual arrival time, in cooperation with the

```
                              supervisor concerned, and the time
                              will be posted, for comparison with the
                              times that the employee clocks in.
                              The idea is not that the employee
                              arrives at an arbitrary hour every day;
                              it is simply that the employee
                              determines the best time for every day
                              to begin.
```

```
        Flex-time, I believe, will improve productivity,
        strengthen employee relations, and provide the firm with
        some very important goodwill among the employees.
        Additionally, it will provide the company with a staffed
        office for a longer period of time each day, so that we
        can be more accessible to our customers. I think that we
        should implement this without delay, and I am more than
        willing to meet with you to discuss any questions you may
        have.
```

External Proposals

External proposals attempt to persuade a potential client or foundation to select and pay for the service or product you offer. They are, therefore, basically sales proposals aimed at marketing your organization's product or service. Sales proposals are either solicited or unsolicited. They range from page-length to book-length. They may be the work of one writer or of a team of writers. Their purpose may be to sell a cleaning service or to bid on the development of a missile system. The destiny of an entire company may rest on a single proposal to the government that took months of preparation and involved contributions from hundreds of people.

Such earth-shaking efforts lie outside the scope of this book. They are usually the work of trained professionals whose job is to write proposals. We will concentrate here on short external proposals, the kind any business person might find it necessary to write.

First, let's distinguish between *solicited* and *unsolicited* proposals. The government or a private organization (such as a company or a foundation) asks for bids on a project. In effect, they say, "We have a job that needs doing. If you think you are best qualified to do it, send us your proposal explaining why." They publish such "requests for proposals" in appropriate specialized publications, or they send the request to companies they know would be able and interested. The "request for proposals" may be quite specific about the format and content of the proposal; if it is, it must be followed precisely. Some agencies and companies refuse to consider proposals that deviate from the required form. Unsolicited proposals differ from solicited proposals in two ways:

1. Instead of demonstrating ability to meet a need the audience has recognized, the writer of an unsolicited sales proposal must persuade the reader that a need or problem exists and that his company has a solution;
2. The writer has more leeway in the format and organization of the proposal, since there are no guidelines.

Your PAFEO process for an unsolicited proposal might look something like this:

Purpose: My purpose is to persuade the customer that there is a problem to which I have a cost-effective solution, so that the customer will decide to buy my service or product.

Audience: I need to determine who our competitors are so I can show the customer how our product has advantages and can benefit this customer more. Would this audience be more interested in quality of personnel, relative cost, turnaround time, convenience?

Format: If the proposal is more than three pages, I'll use an executive summary at the beginning. I want to think how I can make the headings in the body of the proposal draw attention to the matters of greatest interest to the audience.

Evidence: I need to itemize exactly what I will provide, how I will proceed, and when I expect to start and finish each stage. I also need to get precise information on costs. I also need some information on the qualifications of key people.

Organization: I'll use a problem-solution pattern. In the introduction, I'll show the customer the problem and summarize my plan for solving it. In the body of the proposal, I'll include an overview, and then get more specific by using such headings as *Management Plan, Materials, Budget,* and *Personnel.* In the conclusion, I'll summarize and also stress advantages and benefits.

Exercises

1. Read the following sample proposal carefully. It was prepared by a chapter of Women Organized Against Rape (WOAR) and submitted to a foundation for funding.[3] Discuss its effectiveness as a persuasive message:

   · Is it a solicited or unsolicited proposal?
   · How does that affect the content? the organization? the format?
   · Why is the section headed *Description of Agency* longer than the section headed *The Problem*?
   · What differences do you see between the introduction to this proposal and the one we quoted from the English Department in this chapter?
   · Does this proposal have a conclusion? Discuss its content or possible reasons for its omission.

---

[3] Written by Karen Kulp, Women Organized Against Rape, Philadelphia, PA; reprinted by permission of WOAR, Philadelphia, PA.

Women Organized Against Rape

Rape and the Elderly: A Training and Prevention Project

Submitted to: Philadelphia Corporation for the Aging

I. SUMMARY OF PROPOSAL

WOAR requests $8,456 from the Philadelphia Corporation for the Aging to support activities that fit into categories 3, 4, & 5 of the RFP [Request for Proposal], and to promote self-sufficiency, physical and mental health of older people, enhance community supports, and provide support for caregivers of the elderly.

Sexual assault of the elderly is a seldom addressed issue. Misconceptions about rape and its victims lead to elderly women not reporting rape and in inexperience in dealing with sexual assault by professionals who work with the elderly. Even though rape is vastly underreported among the elderly, seven percent of the victims that WOAR saw in the emergency room last year were over sixty. The elderly are more afraid of crime than any other age group.

WOAR proposes to address this issue in three ways:

A. training professionals who work with the elderly on sexual assault awareness and prevention.
B. presenting prevention workshops to elderly populations.
C. developing a brochure about sexual assault and its prevention for older people.

II. DESCRIPTION OF AGENCY

WOAR has provided comprehensive services to sexual assault victims, their families and Philadelphians for ten years, including: a twenty-four hour hotline; accompaniment to the hospital and court for sexual assault victims; advocacy for victims with medical, mental health, and legal institutions; training for professionals about the trauma of sexual assault and care of its victims; and education about sexual assault and its prevention to the general public.

WOAR was the first rape crisis center in Pennsylvania, and is the only center in Philadelphia, serving a population of 1.6 million people in a county which comprises 130 square miles. Since 1973, WOAR has come in contact with close to 17,000 victims and presented approximately 1,800 training or educational programs to 90,000 people.

WOAR services are provided by a combination of volunteer and paid staff, all of whom receive special training. They are recruited from a variety of economic, racial, and cultural backgrounds. Three years ago, WOAR formed a group of volunteers to support older women as volunteers in the organization and to ensure quality services by WOAR to the elderly.

WOAR collaborates with groups concerned with crime and the elderly on policy and legislative improvements. WOAR is an active participant in the Philadelphia Coalition on Crime and the Elderly, which includes a number of different groups that work with the elderly. WOAR's Executive Director is chair of the Women's Caucus, and has been a speaker at their conference for the last two years. WOAR also works regularly with Action Alliance for Senior Citizens. WOAR helped Action Alliance design its court accompaniment program and advised them on dealing with the criminal justice system.

WOAR's educational efforts target groups that are particularly vulnerable. The Child Sexual Abuse Program is an example of a program that WOAR started to meet a community need. Since its inception in 1980, the Child Sexual Abuse Program has provided services for child sexual abuse victims and developed a Prevention through Education Program that is being presented all over the city. In 1982, WOAR saw 492 child sexual abuse victims and presented 116 child sexual abuse prevention programs.

On a smaller scale, WOAR undertook a project this year to reach more members of the Black and Hispanic communities. WOAR trained workers in social service agencies in the Black and Hispanic communities about rape, child sexual abuse, and their prevention. WOAR then led workshops about these issues with the communities served by the agencies. We found that approaching agencies in the community first made for more effective programs and built a base of support and expertise in the community.

III. THE PROBLEM

The elderly are often targets of personal and sexual assaults. In 1982, 7% of the victims that WOAR saw in the emergency room were over sixty years old, including: one 91-year-old, five 90-year-old, one 85-year-old, one 83-year-old, four 82-year-old, two 81-year-old, and seven 80-year-old victims.

Older people often call WOAR's hotline with their concerns. It has been our experience that the elderly are more afraid of crime than any other age group. The fear of crime may cause older people to become reclusive, depressed, paranoid and even more vulnerable.

The belief that rape is a sexually motivated act perpetrated against young beautiful women is false. The reality is that rape is motivated by anger and hostility. People of all ages, races, classes, physical types, and sexes are victimized.

These false beliefs lead to several occurrences. First, an older woman may be reluctant to report a rape, because of her own misconceptions about rape, her shame about it, and her difficulty in describing the actual incident. Second, professionals who work with the elderly may not be looking for signs of sexual assault and may not know how to deal with sexual assault when it is reported. In a recent series of rape-murders in a Philadelphia nursing home, the victims were not even checked for evidence of sexual assault.

IV. PROJECT DESCRIPTION

WOAR intends to address the stated problems in the following manner:

A. *Training for workers in nursing homes and senior centers.*

WOAR proposes to offer a 6-hour comprehensive training to counseling and social service professionals who work with the elderly. The training would have two major goals: 1) to dispel myths and to promote awareness of sexual assault; 2) to give guidelines, techniques, and suggestions for recognizing sexual assault, determining the specifics of its occurrence, and providing quality care for victims in its aftermath.

WOAR has already been asked by the Philadelphia Long-Term Care Ombudsman Program to train their staff. WOAR will develop a training model and pilot it on their staff. After evaluation of the pilot, we will approach nursing homes and senior centers with the training. For the period of this grant, we will offer the training to a total of five nursing homes and senior centers, not including the pilot training. The trainings would be done on site for up to sixteen staff at a time. WOAR's Associate Director Marta Zehner, who is a clinical social worker, and Gola Tatum, WOAR's Training and Education Coordinator, will conduct the training.

After the initial grant, WOAR could contract with individual nursing homes for the training. It is also possible that in the future the program could be funded through WOAR's contract with the City Health Department.

B. *Offering personal safety workshops for senior citizens.*
WOAR has already developed a rape prevention workshop that we use with general audiences (see attachments for agenda). It has been presented to older groups and has been received enthusiastically. We will further examine the content and gear it specifically to the elderly.

WOAR has found that people who participate in the workshop feel more empowered in their lives. One way to begin to alleviate fear is to talk about it openly. This workshop would help people talk about their fear and think about steps they can take to feel less afraid and more in control in their environment. Participants will be asked to evaluate the value of the workshops. WOAR will offer 5 two-hour workshops in nursing homes and senior centers. WOAR will help sponsors develop publicity for the workshop. WOAR has a core of prevention workshop leaders who would facilitate the workshops.

C. *Developing a brochure about rape and its prevention for older people.*
WOAR proposes to develop a brochure about sexual assault for older people. Presently, we have brochures about prevention directed to children, teenagers, and the general population.

The brochure will be written by WOAR's Public Information Coordinator, Karen Kulp, with input from Marta Zehner and Gola Tatum.

It will include general information about sexual assault: what rape is, who gets raped, who rapes, what motivates rapists, and tips for prevention.

WOAR will print brochures in easy-to-read type and will distribute them through senior centers, senior residences, and nursing homes.

2. You are a senior editor of the yearbook and concerned about your successor. You have a certain junior staff member in mind, but that able person has not shown much interest. Write a persuasive letter that will sell that student on taking the job. Review Ewing's Eleven Rules of Persuasion before you write.

3. Here is a letter from a volunteer fire company soliciting contributions from the townspeople. Analyze it as an example of the use of the AIDA formula. Assume that you are chief of the same company and now know the importance of AIDA. Write a sales letter that will get action. Start with the PAFEO process and then use AIDA as a way of attaining your purpose.

Dear Property Owner,

First of all we, on behalf of the members of your Volunteer Fire Department, want to express our sincere appreciation for the generous return to our 1986 appeal

for voluntary contributions. It was an inspiration to the
firemen and greatly helps to strengthen their Aid Fund and
assist them in Community projects.

Your Fire Department strives for "maximum fire
protection--minimum fire loss." This helps to keep our
fire insurance rates in a low bracket. You could help us
by installing smoke alarms in your home for protection.

We pledge our best for your protection and will be
pleased to have your support.

Sincerely,

4. You are an assistant director of admissions at your college or university. Your
   boss, the director, asks you to draft a letter to be sent to high school seniors in
   your state with CEEB scores totaling over 1000. The action line is to get them
   to call the Admissions Office to schedule a visit to the campus. Catalogs,
   viewbooks, and admissions brochures should provide you with information you
   can use as evidence in the sales letter you will write.

5. The OneYear Tire Company lists these Adjustment Policy Limitations in its Limited
   Warranty.

   The following are not covered by this policy:
   Irregular wear or tire damage due to
   —road hazards
   —wreck, collision, or fire
   —improper inflation, negligence, racing, chain damage
   —mechanical condition of the vehicle

   Your radial tires, guaranteed for 40,000 miles, are dangerously worn after only
   10,000 miles. You had a wheel alignment when you purchased the tires a year
   ago; the wheels have not been aligned since, and the dealer tells you the irregular
   wear is the result of alignment problems caused by potholes and winter roads.
   He points out that the warranty does not cover tire problems caused by the
   mechanical condition of the car.
   You decide to request an adjustment from the OneYear Tire Co., your argument
   being that the warranty says nothing about wheel alignments being required at
   intervals under 10,000 miles. You claim you acted in good faith in properly
   maintaining your car, and you request a credit of 50 percent of the $135.00 you
   originally paid for the tires. Address your letter to Mr. Frank Johnson, Customer
   Service Representative, OneYear Tire Co., Minneapolis, Minnesota.

6. When your class has finished writing these requests for adjustments, exchange
   your letter with a classmate. In the role of Frank Johnson, answer your classmate's
   letter. You must tell the writer that no adjustment will be made because the
   excessive tire wear was caused by the condition of the vehicle, not by any defect
   in the tires themselves. Follow the indirect approach for bad news letters, that
   we discussed in Chapter 7. Be sure to indicate that you fully understand the

complaint and are sympathetic to the customer, outline the reasons or policy behind your decision, and give the bad news. End with a statement that will maintain as much goodwill as possible.

7. Write an informal proposal to the Director of Food Services proposing some change in that operation.

8. Write a request to your department head that you be allowed to substitute one course of your choice for a course that is always required of majors in your field.

9. Write a proposal to your Security or Police Department that will offer a solution to a particular parking problem on your campus or in your town.

CHAPTER 9

Visuals

Whether you want to inform or persuade, visuals can help you achieve your purpose and meet the needs of your audience. Visuals are simply another way to organize your evidence—especially statistical and comparative data—into a format accessible to your readers.

Visuals and the PAFEO Process: An Overview

Purpose

Whether your visuals are hand-produced or computer-generated, visuals have several purposes:

They present complicated data in a form that can be more easily understood.
They enhance and make more readable the presentation of any data that are paired with prose.
They condense information.

They allow readers to pull together or sort through diverse data to make comparisons and contrasts more immediately.[1]

Audience

The right visual can help save your busy reader time. Moreover, certain readers, especially those with a technical or statistical background, may rely more on the visuals than the text of a document for the information they need. The key in using visuals effectively is matching specific visuals to specific audience needs.

Format

Applying the concept of format to the use of visuals means making a choice among types of visuals. Visuals include tables, graphs, charts, diagrams, and photographs.

Evidence

Effective use of visuals requires that you match your choice of visual to the data you have at hand. Tables, for instance, present exact data, while graphs show trends, comparisons, changes, or relationships between variables. Charts also illustrate relationships, but they are not plotted on a system of coordinates. The most common types are pie charts, flow charts, and organizational charts. Diagrams or illustrations show the parts of an item or the steps in process, while photographs provide an exact visual record of an event or item.

Organization

Visuals can be organized in relationship to the text of a document in two ways. When there is a close integration between text and visual, both should appear in the body of the document as close to each other as possible. Explanatory text should, however, always precede the visual, so that your readers see what you intend them to see in the visual.

When your visuals are secondary to your text, you can place them together in an appendix to your document.

Remember at all times, however, to maintain a balance between text and visuals. Too many visuals can cause your readers to lose contact with your train of thought. Moreover, overuse of visuals diminishes their impact and defeats your very purpose in using them.

[1] Steven Darian pointed out that some experiments suggest that we remember materials presented by graphic aid longer than materials simply presented in prose. See his article, "Using Algorithms, Prose, and Graphics for Presenting Technical and Business Information," *ABCA Bulletin,* 46 (December, 1983), 26.

Matching Visuals to Data: A Brief Case Study

Visuals are time-consuming items to produce, unless you have access to an art department or to a computer graphics system. Given the appropriate software, you can create and make decisions about the appropriateness of individual visuals to the data you want to present.

In the balance of this chapter, we will discuss in greater detail the kinds of visuals available to you. For present purposes, however, let's see how a software program works not only to produce a visual but also to help a writer choose the visual most appropriate to the data being presented.

Suppose you wanted to present data on ABC Company's sales, costs, and gross profits over two years, supplying figures in these three areas by the quarter.

First, you could use the software to generate a simple *table* (see Table 9–1). Next, using the same software, you could produce a *line graph* to show the data (see Figure 9–1).

Third, again using the same software, you could present your data in a *bar graph*. Actually, you could present the data in two different kinds of bar graphs. You could use a *segmented bar graph* (see Figure 9–2), or a *multiple bar graph* (see Figure 9–3).

Finally, you could use the same software to present your data in a series of three *pie charts* (see Figure 9–4 for a pie chart showing quarterly sales).

In presenting these data visually, the table is, from the reader's point of view, the least helpful. As a general rule, tables are too time-consuming to read. Table 9–1 *is* short, but like all tables, it does not allow for easy comparisons. It forces the reader to work harder than is necessary at understanding the data presented.

The pie chart is also not of much use to the reader, since the sizes of the segments of the pie are not sufficiently and significantly different to show any clear comparison among data.

TABLE 9–1   Quarterly Sales, Costs, and Gross Profits

A B C COMPANY

QUARTER	SALES	COST	GROSS PROFIT
1	93420	49368	44052
2	90650	48260	42390
3	99780	51912	47868
4	102440	52976	49464
5	110990	56396	54594
6	125300	62120	63180
7	95220	50088	45132
8	75930	42372	33558

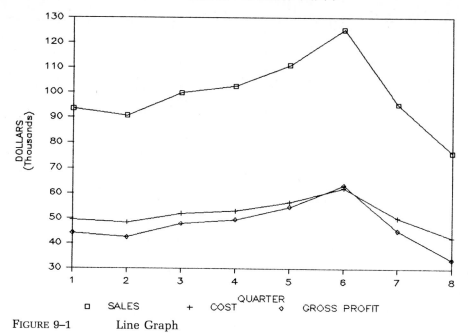

ABC COMPANY

FIGURE 9–1   Line Graph

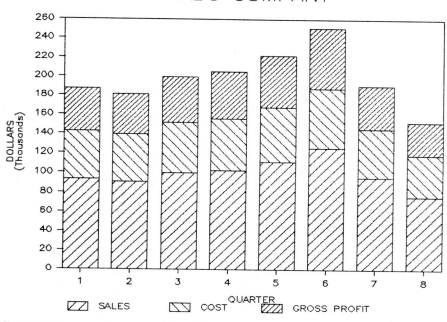

ABC COMPANY

FIGURE 9–2   Segmented Bar Graph

196

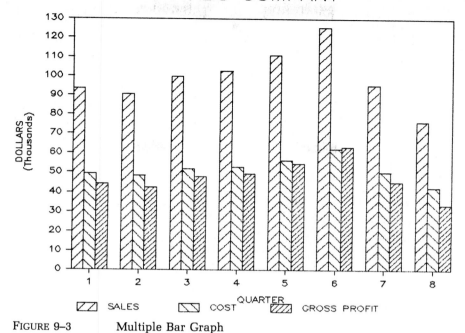

ABC COMPANY

FIGURE 9–3   Multiple Bar Graph

ABC COMPANY

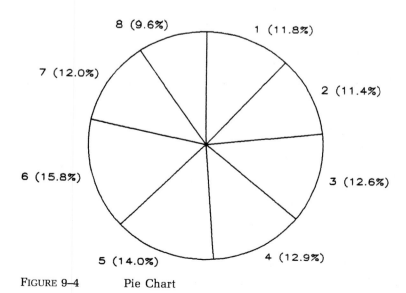

FIGURE 9–4   Pie Chart

The line graph does show all three sets of data clearly, but more for individual than comparative purposes. The reader needs the visual both to see the data and to interpret them.

The two bar graphs prove most helpful to the reader. Both allow the reader to see and to compare the data simultaneously. The segmented bar graph, we would argue, does so better than the multiple bar graph. The segmented bar graph allows the reader immediate access to three sets of data for each quarter and best helps the writer communicate his evidence to a busy reader.

Tables

Consider the following paragraph from an article on a 1982 movement by black students at the Harvard Law School to boycott a course on race discrimination because one of the instructors was white. The paragraph lists the results of recent bar examinations.

> The statistics that show that members of some minority groups—particularly blacks—have scored considerably lower than whites on bar examinations are disturbing. For example, of 5,556 first-time takers of the fall, 1978, California bar exam, 67 per cent of the whites and 53.4 per cent of the Asians, but only 30.9 per cent of the Hispanics and 19 per cent of the blacks, passed. Of 5,159 who took the fall, 1980, exam for the first time, 69 per cent of the whites passed; for the Asians the figure was 57 per cent, for Hispanics 44.1 per cent, for blacks 31.4 per cent. When 4,889 took the fall, 1981, exam for the first time, 67.7 per cent of the whites, 58.7 per cent of the Asians, 37.9 per cent of the Hispanics, and 31.2 per cent of the blacks passed. The spring, 1982, data show that of 1,373 first-timers, 49.1 per cent of the whites and 33.3 per cent of the Asians, but only 23.1 per cent of the Hispanics and 11.1 per cent of the blacks, were successful.[2]

Understand all that? Probably not on the first, the second, or the third reading. As a reader, you are simply overwhelmed by number after number, and the writer's purpose, to inform his readers of significant data, is stymied.

What was needed here was not straight horizontal prose, but rather a simple vertical table. Such a table would have allowed the reader to grasp data and make any necessary comparisons easily.

Table 9–2 presents the same information in a simple vertical table. Notice how accessible the information previously buried in a paragraph of prose has now become.

Tables are best suited for the presentation of detailed, specific data. Tables give readers exact figures or information and allow readers to see relationships clearly between exact figures and information.

Table 9–3 appears in a package containing information and forms for

[2] John H. Bunzel, " 'Flawed Assumptions' Behind the Boycott by Black Students at Harvard Law School," *The Chronicle of Higher Education,* 25 (October 27, 1982), 64.

TABLE 9–2 Percentages by Race of the First-time Takers of the California Bar Examination Who Passed

Date of the Examination	Total Number Taking the Examination for the First Time	Percentage Passing			
		White	Asian	Hispanic	Black
Fall 1978	5,556	67.0%	53.4%	30.9%	19.0%
Fall 1980	5,159	69.0	57.0	44.1	31.4
Fall 1981	4,889	67.7	58.7	37.9	31.2
Spring 1982	1,373	49.1	33.3	23.1	11.1

TABLE 9–3 1985 Tax Table

If line 37 (taxable income) is—		And you are—			
At least	But less than	Single	Married filing jointly *	Married filing separately	Head of a household
		Your tax is—			
$0	$1,775	$0	$0	$0	$0
1,775	1,800	0	0	2	0
1,800	1,825	0	0	5	0
1,825	1,850	0	0	7	0
1,850	1,875	0	0	10	0
1,875	1,900	0	0	13	0
1,900	1,925	0	0	16	0
1,925	1,950	0	0	18	0
1,950	1,975	0	0	21	0
1,975	2,000	0	0	24	0
2,000					
2,000	2,025	0	0	27	0
2,025	2,050	0	0	29	0
2,050	2,075	0	0	32	0
2,075	2,100	0	0	35	0
2,100	2,125	0	0	38	0
2,125	2,150	0	0	40	0
2,150	2,175	0	0	43	0
2,175	2,200	0	0	46	0
2,200	2,225	0	0	49	0
2,225	2,250	0	0	51	0
2,250	2,275	0	0	54	0
2,275	2,300	0	0	57	0
2,300	2,325	0	0	60	0
2,325	2,350	0	0	62	0
2,350	2,375	0	0	65	0
2,375	2,400	0	0	68	0

If line 37 (taxable income) is—		And you are—			
At least	But less than	Single	Married filing jointly *	Married filing separately	Head of a household
		Your tax is—			
2,400	2,425	2	0	71	2
2,425	2,450	5	0	73	5
2,450	2,475	8	0	76	8
2,475	2,500	11	0	79	11
2,500	2,525	13	0	82	13
2,525	2,550	16	0	84	16
2,550	2,575	19	0	87	19
2,575	2,600	22	0	90	22
2,600	2,625	24	0	93	24
2,625	2,650	27	0	95	27
2,650	2,675	30	0	98	30
2,675	2,700	33	0	101	33
2,700	2,725	35	0	104	35
2,725	2,750	38	0	106	38
2,750	2,775	41	0	109	41
2,775	2,800	44	0	112	44
2,800	2,825	46	0	115	46
2,825	2,850	49	0	117	49
2,850	2,875	52	0	120	52
2,875	2,900	55	0	123	55
2,900	2,925	57	0	126	57
2,925	2,950	60	0	129	60
2,950	2,975	63	0	132	63
2,975	3,000	66	0	135	66
3,000					
3,000	3,050	70	0	140	70
3,050	3,100	75	0	146	75
3,100	3,150	81	0	152	81
3,150	3,200	86	0	158	86
3,200	3,250	92	0	164	92
3,250	3,300	97	0	170	97
3,300	3,350	103	0	176	103
3,350	3,400	108	0	182	108

If line 37 (taxable income) is—		And you are—			
At least	But less than	Single	Married filing jointly *	Married filing separately	Head of a household
		Your tax is—			
3,400	3,450	114	0	188	114
3,450	3,500	119	0	194	119
3,500	3,550	125	0	200	125
3,550	3,600	131	4	206	130
3,600	3,650	137	9	212	136
3,650	3,700	143	15	218	141
3,700	3,750	149	20	224	147
3,750	3,800	155	26	230	152
3,800	3,850	161	31	236	158
3,850	3,900	167	37	242	163
3,900	3,950	173	42	248	169
3,950	4,000	179	48	254	174
4,000					
4,000	4,050	185	53	261	180
4,050	4,100	191	59	268	185
4,100	4,150	197	64	275	191
4,150	4,200	203	70	282	196
4,200	4,250	209	75	289	202
4,250	4,300	215	81	296	207
4,300	4,350	221	86	303	213
4,350	4,400	227	92	310	218
4,400	4,450	233	97	317	224
4,450	4,500	239	103	324	229
4,500	4,550	245	108	331	235
4,550	4,600	251	114	338	241
4,600	4,650	258	119	345	246
4,650	4,700	265	125	352	252
4,700	4,750	272	130	359	258
4,750	4,800	279	136	366	264
4,800	4,850	286	141	373	270
4,850	4,900	293	147	380	276
4,900	4,950	300	152	387	282
4,950	5,000	307	158	394	288

* This column must also be used by a qualifying widow(er).

Source: The Internal Revenue Service.

TABLE 9–4   Comparison of Health Insurance and Health Maintenance Options

Benefit Plan Name	Blue Cross/Blue Shield 100/ Major Medical	Philadelphia Health Plan	Health Service Plan of Penna.
Abbreviation	*BC/BS/MM*	*P.H.P.*	*H.S.P.*
Type of Plan	Hospitalization, medical, surgical service and major medical insurance	*GPP	*GPP
Service Area and Emergencies	Guaranteed benefits in any approved hospital—services of any physician up to U.C.R.**	5 Delaware Valley Counties and Burlington, Camden and Gloucester Counties—Emergency treatment anywhere covered in full	5 Delaware Valley Counties in PA, plus parts of NJ Emergency treatment anywhere—Covered in full
Inpatient Hospital	Up to 120 days, semi-private room with $5/day co-payment for first 10 days	No maximum limit	No maximum limit
Outpatient Treatment	Covered at hospital within 72 hours of accident or med. emergency—$5 co-payment	Covered in full	Covered in full
Physician Visits:			
Hospital	Covered in full	Covered in full	Covered in full
Office	80% covered Maj Med, $100 deducted	Covered in full	Covered in full
Home	21 visits covered if applicant subscriber is totally disabled, $25 deducted	Covered in full	Covered in full
Physician Care:			
Surgery Anesthesia	Covered up to UCR**	Covered in full	Covered in full
Consultants	Covered up to UCR**	Covered in full	Covered in full

taxpayers who file Form 1040 in completing their federal income tax. The data appear in vertical columns that allow the taxpayer easy access to needed information.

The format and organization of a table are important. First, make sure you label your table with a specific title. Columns within a table need equally specific titles. In Table 9–3, white space and rules, solid or broken lines, are used to distinguish the columns and to separate the income levels. Rules are also used to separate table headings from data.

TABLE 9–4   (*continued*)

Physician Care (cont.)			
Laboratory X-ray Cardiogram	Covered in full for diagnostic purposes only	Covered in full	Covered in full
Maternity	Covered up to UCR**	Covered in full	Covered in full
Preventive Medicine, Physical Exam	Not Covered	Covered in full	Covered in full
Eye, Ear Exams	Not Covered	Covered including refractions. Hearing aids and glasses not covered	Covered including refractions. Hearing aids and glasses not covered
Mental Inpatient: Hospital Physician	Coverage for 30 days. After 365 days, the 30 days are renewed	30 days per Benefit Year 30 days per Benefit Year	45 days per year 45 days per year
Mental Outpatient: Physician	Up to 50 visits per year, $12.50 per visit	20 visits per year—first 3 visits covered in full. Next 17—co-pay $10 per visit	30 visits per year—first 3 visits covered in full. Next 27—co-pay $10 per visit
Dependent Definition	Spouse and unmarried dependent children to age 19 or to age 23 if a full-time student	Spouse and unmarried children to age 19 or to age 23 if a full-time student	Spouse and unmarried dependent children to age 19 or to age 23 if a full-time student
Premium Rates:	Rates will be available during open enrollment period	Rates will be available during open enrollment period	Rates will be available during open enrollment period

* Prepaid Group Practice Plan (GPP)—a team of personal physicians and medical professionals practice together to provide members with medical care in a multispecialty medical center.
** Usual, customary, or reasonable.

Source: Take Account of Your Health, University of Pennsylvania, 1984. Reprinted with permission. University of Pennsylvania Department of Human Resources/Benefits, Philadelphia, PA.

To avoid confusion in explanatory footnotes, especially with statistical tables, indicate footnotes with lowercase letters or with symbols such as asterisks, stars (*, **, ***), or daggers (†, ‡, ‡) when lower case letters would confuse readers.[3]

When appropriate, use a source line at the end of the table, either to one side or centered, identifying the source of the information. To avoid confusion and unnecessary page turning, run your data in only one direction

---

[3] *The Chicago Manual of Style,* 13th ed. rev. (Chicago: University of Chicago Press, 1982), 12.49, suggests the following order for symbols indicating footnotes in tables: * (asterisk or star), † (dagger), ‡ (double dagger), § (section mark), ‖ (parallels), and # (number sign). When necessary, each symbol can then be doubled or tripled.

in any table. In horizontal tables, vertical information has no place, and vice versa.

If you have several tables, number them sequentially, refer to them by their actual numbers in your text (use "Table 1" rather than "the table below"), and include a separate initial listing of tables by number, title, and page in any document with more than five tables.

You can use tables to present other than statistical data, and tables will be essential if you are presenting complicated data that you expect your reader to understand, analyze, or compare. Table 9–4 (pages 200–201) is probably the only easy way to inform employees about their medical insurance options. The alternative would be to use a lengthy booklet that no one would want to read.

Graphs

We see and use tables often—bus, train, and plane schedules, the monthly telephone bill, the report of the daily trading activities on any of the various stock exchanges, the roster of courses being offered in the next semester. Tables are any easy way for readers to get at exact pieces or sets of data.

Graphs, on the other hand, show trends, cycles, shifts, or distributions in data. Graphs come in two forms—*line graphs* and *bar graphs.* Both types of graphs are plotted on a system of coordinates designed to show the relationship between two sets of variables.

Line Graphs

Line graphs show trends or the interaction between two variables. When showing trends, use the vertical axis to indicate amounts and the horizontal axis to show time or a quantity being measured. Both axes must proceed in equal increments. The vertical axis should be scaled to begin at zero or indicate with a break that some of the increments have been left out.

Figure 9–5 represents a simple line graph. Figure 9–6 displays the same data as Figure 9–5, but, because the vertical axis doesn't begin at zero or indicate the zero point, the graph distorts the data. Such a graph is a *truncated graph,* one of many ways a writer can lie with statistics.[4] We assume that, when you use visuals, you will be using them to enlighten, not to conceal.

You can use line graphs to compare several sets of data for the same period of time, as long as the separate lines of a *multiple graph* remain distinct and the graph itself remains uncluttered. A multiple line graph should identify each separate line with a legend or a label on the graph itself.

To add to the usefulness of either simple or multiple line graphs, you

---

[4] The classic study of how statistics can be misused to make the insignificant seem important remains Darrell Huff's *How to Lie with Statistics* (New York: Norton, 1954).

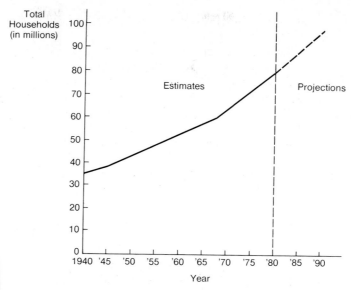

FIGURE 9–5   Estimated Number of Households, 1940–
               1980, with Projections to 1990
               Source: Federal National Mortgage Association.

can indicate specific values for the points on the line(s) of the graph. Figure 9–7 illustrates a multiple line graph that also indicates the value of each of the points on the line graph.

To guarantee that your line graphs are effective, label them with specific titles. The title can be centered or aligned with the left margin under the graph.

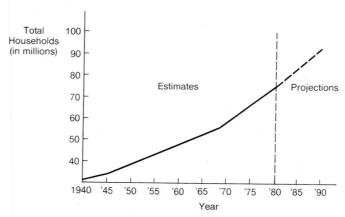

FIGURE 9–6   Data from Figure 9–5 Inaccurately Pre-
               sented in Truncated Form

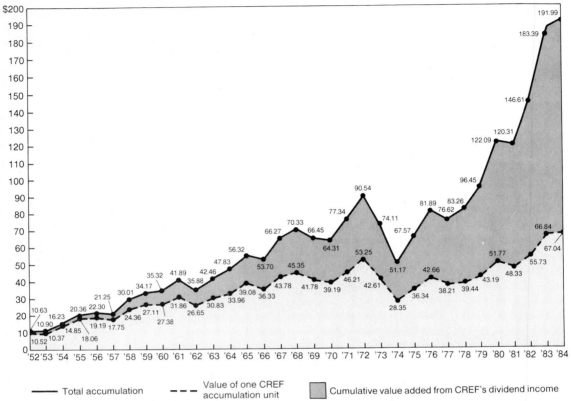

| | Total accumulation | | Value of one CREF accumulation unit | | Cumulative value added from CREF's dividend income |

FIGURE 9–7

Accumulation Resulting from Purchase of One CREF Accumulation Unit in July 1952 (All Values Shown at December 31)

Note: This graph shows CREF's past investment experience and should not be considered a prediction of CREF's performance for the years ahead.

Source: Teachers Insurance and Annuity Association/College Retirement Equities Fund. Reproduced by permission.

When a document contains multiple graphs, assign each a number in sequential order (Figure 1, Figure 2, and so forth). When a document contains more than five graphs, list them by number, title, and page on a separate page following the table of contents at the beginning of the document. All lettering on and around graphs should be horizontal. Place source lines and footnotes, where appropriate, below the title of the graph on the left. Place the caption for the vertical axis, again where appropriate, at the upper left. When producing graphs on graph paper, be careful to keep the grid lines faint so that the curve representing the data stands out.

Bar Graphs

Bar graphs display data in horizontal or vertical bars of equal width. The length of the bars is scaled to represent different quantities. Bar graphs can be simple, multiple, or segmented.

Use a *simple bar graph* such as Figure 9–8 to show changes in one item over time. Use *multiple bar graphs* to compare the size of different items at one time, as Figures 9–9 and 9–10 do vertically.

Use a *multiple bar graph* to indicate the relationship among several items over time. Limit multiple bar graphs to three bars at each major point on the vertical or horizontal axis, and include a legend explaining the meaning of the several bars. Figure 9–11 is an example of an effective three-bar graph.

Segmented bar graphs show the relative size of the individual parts of a whole. Segmented bar graphs are similar to pie charts, only they allow you to show a greater number of the parts of a whole than pie charts do. Figure 9–12 uses three segmented bar graphs, one each for 1981, 1982, and 1983, to indicate an increase in revenue during those years.

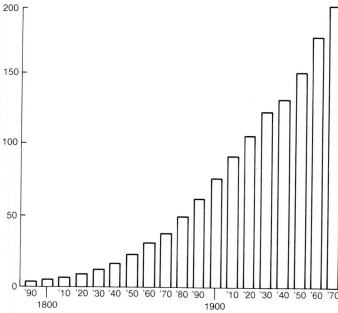

FIGURE 9–8   Population of the United States: Total Number of Persons in Each Census (1790–1970)
Note: Numbers are in millions.
Source: U.S. Census Bureau.

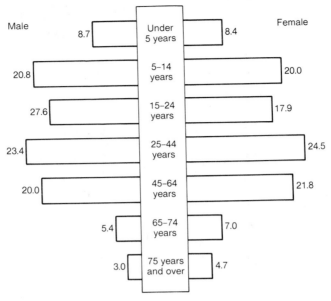

FIGURE 9–9   Number of Persons by Age and Sex: 1970
Note: Numbers are in millions.
Source: U.S. Census Bureau.

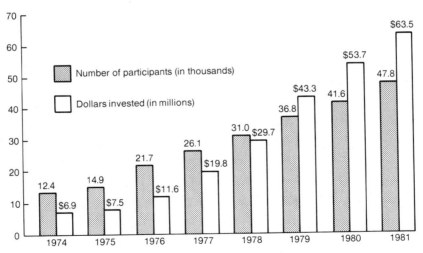

FIGURE 9–10   Dividend Reinvestment Plan Participation
Source: Commonwealth Edison 1981 Annual Report. Reprinted by permission.

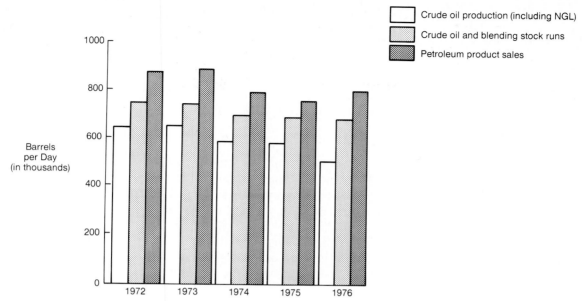

FIGURE 9–11   Crude Oil Sufficiency—Production Versus Refinery Runs and Petroleum Product
Sales, 1972–1976
Source: Atlantic Richfield Company Annual Report 1976. Reproduced by permission.

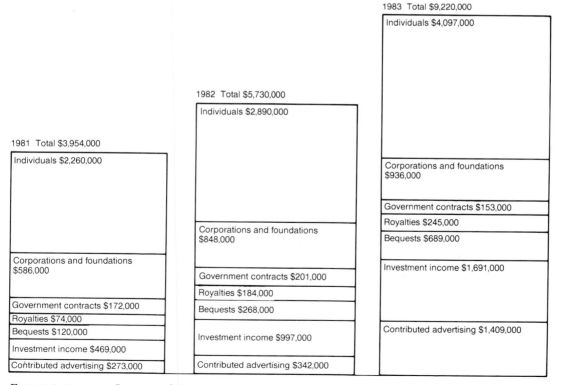

FIGURE 9–12   Sources of Revenue
Source: World Wildlife Fund. Reprinted by permission.

Charts

Charts show numerical, physical, or conceptual relationships. Three of the most commonly used charts in business and industry are the pie chart, the organizational chart, and the flow chart.

Pie Charts

Like segmented bar graphs, *pie charts* show the relationship between parts and the whole. However, pie charts are not as versatile as bar graphs.

Authorities differ on how many pieces of the pie are too many in a pie chart; we prefer using them for visuals where the whole is to be broken up into no more than six parts. Since pie charts present data as wedge-shaped sections of a circle, the complete circle equals the whole of 100 percent of some quantity. Each percentage point is then equal to 3.6°.

In constructing pie charts, begin at the 12 o'clock position and sequence the wedges clockwise beginning with the largest and ending with the smallest. If you shade the wedges, also move clockwise from light to dark. If you use colors for the wedges, use no more than three shades of the same color, and make sure that the shades are visually distinct. Label each wedge, and assign it a percentage value. Keep all text on and around the pie chart horizontal. As a final check, make sure the parts of the pie chart really add up to 100 percent. Figure 9–13 shows a typical pie chart.

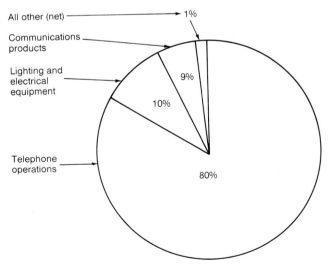

FIGURE 9–13   Source of Combined Income (Continuing Operations) 1974–1978
Source: General Telephone and Electronics Annual Report 1978.
Reprinted by permission of GTE Corporation.

Organizational Charts

Organizational charts show how the various parts or people in an organization are related to one another. They provide a useful overview of the lines of authority within an organization. They can be general, listing only departments and job titles, or specific, indicating who fills each position within the organization.

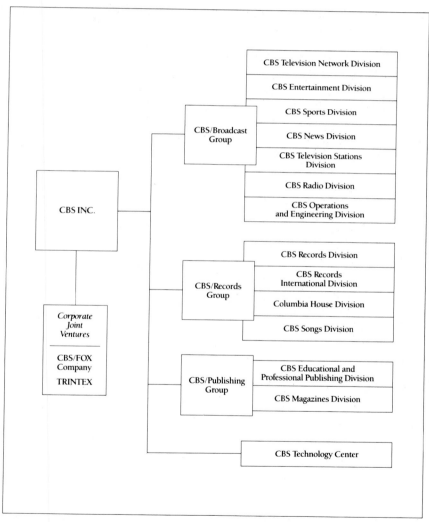

FIGURE 9–14   The CBS Organization
Source: The 1985 CBS Annual Report to Shareholders. Reprinted by permission of CBS, Inc.

You can construct organizational charts horizontally or vertically. In a horizontal chart such as Figure 9–14, levels of authority decrease from left to right. In a vertical chart such as Figure 9–15 levels of authority decrease from top to bottom, while levels of responsibility increase from bottom to top.

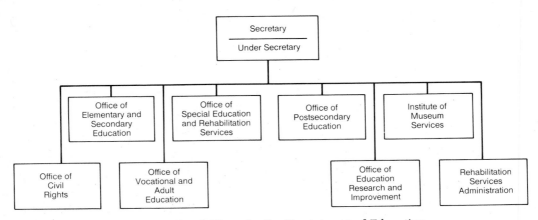

FIGURE 9–15   Organizational Chart for the Department of Education
Source: From INFORMATION U.S.A., by Matthew Lesko. Copyright © 1983 by Matthew Lesko. Reprinted by permission of Viking Penguin, Inc.

Flow Charts

Flow charts illustrate the steps in a process from beginning to end. They allow readers to see at a glance the steps in a procedure. Figure 9–16 shows the steps by which a book moves from manuscript copy to printed form, clearly indicating the steps and the people involved at each step in the procedure.

Diagrams

Diagrams are simply sketches or drawings. They detail the steps in a procedure or show the parts of an item. While diagrams share a purpose with several visuals we have already discussed, they are more difficult—and more costly—to produce. Unless you are a skilled artist, you will need the assistance of your company's art department or the aid of a computer graphics software package to produce diagrams.

Figure 9–17 consists of three diagrams detailing the steps in the procedures for turning oil sands into synthetic crude, shale into oil, and coal into oil or gas. These diagrams accompany an article on synthetic fuels—a fairly technical subject—intended for the layperson. They allow the writer to turn

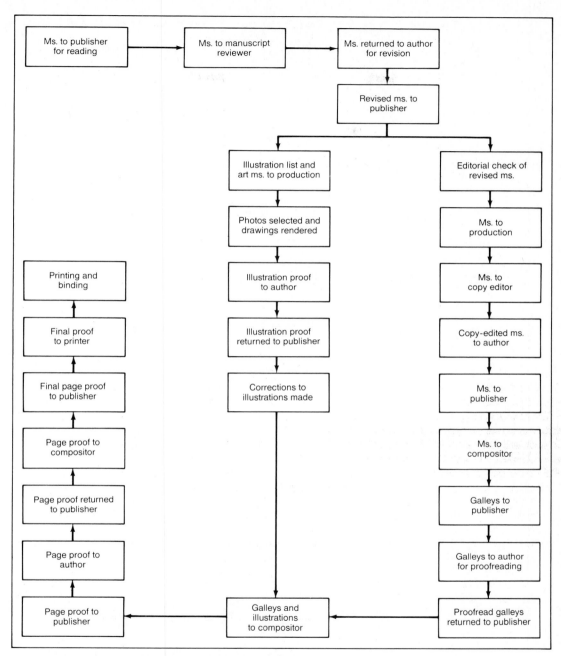

FIGURE 9–16   From Manuscript (Ms.) to Printed and Bound Book

211

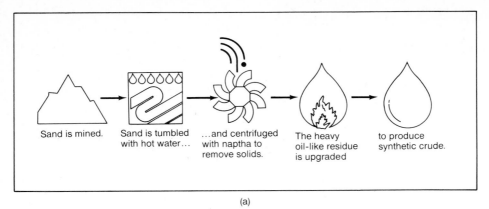

(a)

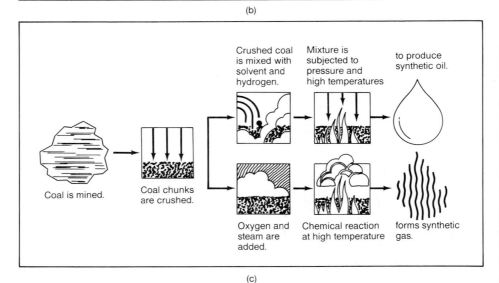

(b)

(c)

FIGURE 9–17   (a) Turning Oil Sands into Synthetic Crude; (b) Turning Shale into Oil; (c) Turning Coal into Oil or Gas
Source: Adapted by permission, from Randall Meyer, "Planning a Synthetic Fuels Industry," *Exxon USA,* 20 (Second Quarter 1981), 21–26.

what could be simply writer-based prose into reader-based prose by letting the reader "see" what the writer is talking about.

Exploded diagrams appear in maintenance manuals or as attachments to sales presentations. They show how the parts of an item are assembled or why one product is better than another. Figure 9–18 shows top, back, and front views of a completely assembled 3M overhead projector.

Figure 9–19 provides exploded diagrams of five parts of the same overhead projector. An accompanying table (Table 9–5) makes it easy for the reader to order replacements for any individual items in these five projector parts.

Pictograms (such as Figure 9–20 on page 216) function like bar graphs but use symbols instead of bars to display data either horizontally or vertically. The symbols used in pictograms must be self-explanatory: use cars for automobile sales, homes for housing starts, coins for money raised or spent, and so forth. Also make sure to include a key explaining what one basic symbol equals. Pictograms will only present approximate data. For more precision, you can add precise figures to a pictogram.

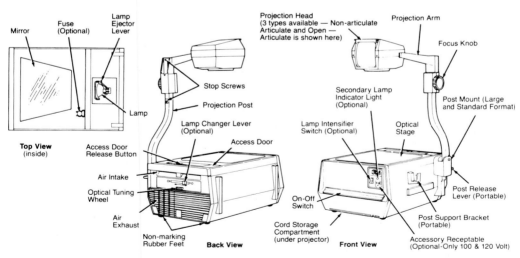

FIGURE 9–18   Getting Acquainted with Your 213 Projector
                     Source: Reprinted by permission of and copyrighted by Minn. Mining and Mfg. Co.

Post mount, arm and focus assembly

Lamp changer/lamp holder assembly

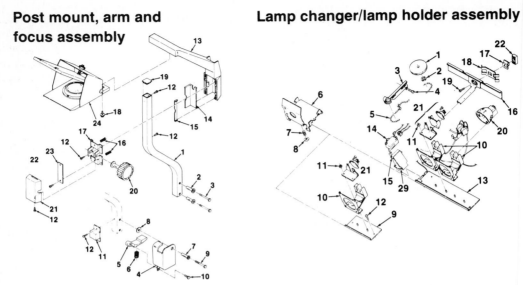

FIGURE 9–19   Parts Identification
Source: Reprinted by permission of and copyrighted by Minn. Mining and Mfg. Co.

TABLE 9–5   List of Parts

Post Mount, Arm, and Focus Assembly

Item No.	Description	Part Number
1	Post	78–8011–1154–9
	Non-Portable	
2	Bushing-Guide	78–8011–1251–3
3	Screw M5 x .8 x 35	26–1000–4860–7
	Portable	
4	Cover Assy., Portable Hinge	78–8011–1430–3
5	Lock	78–8011–1248–9
6	Spring Lock	78–8011–1252–1
7	Bushing—Pivot (Portable)	78–8011–1250–5
8	Washer—Plastic	78–8011–1349–5
9	Screw, M5 x .8 x 40	26–1000–4861–5
10	Screw, M5 x .8 x 14	26–1000–2421–0
11	Shield—Post Locks	78–8011–1422–0
12	Screw, M3 x .5 10	26–1000–6843–1
13	Arm	78–8011–1352–0
14	Wear Pad (Angle)	78–8011–1220–8
15	Wedge	78–8011–1230–7
16	Spring—Wedge	78–8014–8984–6
17	Bracket	78–8014–8983–8
18	Screw—Head Mounting	26–1001–4122–0
19	Plug—Post Cap	78–8011–1420–4

TABLE 9–5 (*continued*)

20	Knob and Roller Assy.	78–8015–2358–6
21	Cover—Roller	78–8011–1308–1
22	Screw 8–16 x 3/8	26–1000–1167–1
23	Spring—Roller Pad	78–8011–1228–1
24	Projection Head Assy.	

Lamp Changer/Lamp Holder Assembly

Item No.	Description	Part Number
1	Knob, Color Tuning	78–8011–1340–4
2	Retaining Ring	78–8656–4001–1
3	Ider—Slide	Not Available
4	Screw M4 x 7 x 8 Taptite	26–1000–6745–8
5	Chain—Bead	78–8011–1227–3
6	Lamp Carriage Assembly	
	Standard Aperature	78–8011–1222–4
	Variable Iris Assembly	Not Available
7	Spring	78–8014–9017–4
8	Washer	Not Available
9	Lampholder Assembly Plate	78–8014–8813–7
	(Single Lamp)	
10	Lampholder	
	36V Dual & Single	78–8015–2309–9
	82V Single	78–8011–1189–5
	82V Dual	78–8014–8810–3
	100V Single	Not Available
11	Nut—M3 x .5	78–8010–7420–0
12	Screw, M3 x .5 x 6 Taptite	26–1000–0338–8
13	Changer Slide Assembly	78–8011–1454–3
14	Switch—Sensing	78–8014–8833–5
15	Insulator—Switch	78–8011–1192–9
16	Yoke—Changer	78–8011–1224–0
17	Stud—Changer	Not Available
18	Detent—Changer	" "
19	Screw, 6–19 x 3/8 PT	26–1000–0336–2
20	Lamp	
	ENX, 360W, 82V	78–8011–1186–1
	ERV, 320W, 36V	78–8011–1403–0
	EPW, 360W, 100V	Not Available
21	Thermostat, Assembly	
	Single Lamp Unit	78–8014–9075–2
	Dual Lamp Unit	78–8011–9074–5
22	Knob—Changer	78–8011–1266–1

Source: Reprinted by permission of and copyrighted by Minn. Mining and Mfg. Co.

Figure 9–20 provides data, as its title indicates, "at a glance" using figures to indicate the number of donors and money bags to indicate the amounts of money given in thousands of dollars.

Pictograms can distort data if they aren't constructed carefully. To show larger quantities on a pictogram, increase the number, not the size, of the symbol used. Figure 9–21 shows one maintenance man almost twice as big

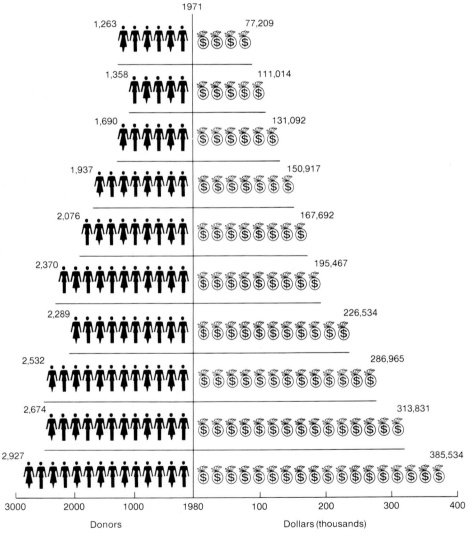

FIGURE 9–20   Wharton Graduate Annual Giving Ten Years at a Glance, 1971 to 1980
Source: Adapted from *Anvil,* The Wharton Graduate Alumni Magazine, 5 (Fall 1980), p. 3 insert. (The artist for the original pictogram was Howard Kline.)

FIGURE 9–21   Our Maintenance Men Get Paid
                     Forty-six Percent More

as the other, but the larger drawing covers an area almost four times as great on the page. "Our" maintenance men simply look like giants; the pictogram doesn't accurately indicate the difference in pay scales. The artist needs to increase the number, not the size, of the figures.

Photographs

Also more complicated and often more expensive to produce than tables and graphs, *photographs* can be used to enhance or take the place of prose. Photographs show the surface appearance of an object, record an event, or document the development of a subject in various stages or over different periods of time.

Diagrams, on the other hand, can show the inner workings of a mechanism. They can desmystify the steps in a process. They can convey complicated information in easy-to-follow form. Diagrams can pick up where photographs leave off.

Photographs like diagrams are, however, best left to the experts in an audiovisual or art department. When in doubt about your own abilities as a photographer or an artist, get assistance from the professionals.

The Integration of Text and Visuals

While it is important to choose the appropriate visuals for the data you wish to present, it is equally important that you position your visuals where they will be most useful to your reader.

Always refer to a visual in the text before using it. It is important to provide your readers with a frame of reference for a visual before plopping it in the middle of a page.

Remember to locate all visuals as closely as possible after their first reference in text. A British study showed that

> when diagrams were positioned so that they immediately followed the sentence in which they were first referenced, then readers spent more time attending to the diagrams and subsequently judged the passage as "clearer" and their own understanding as "greater" than when the diagrams were distributed uniformly throughout the text with no relation to specific content.[5]

Exercises

1. The following data all concern Grenadian nutmeg. The sources for the data are the U. S. Department of Agriculture and the World Bank: In 1981, nutmeg was Grenada's second-largest export. The island's total exports were worth $18.8 million. Cocoa accounted for $7 million; bananas, $3.6 million; nutmeg, $3.8 million; other products, $4.4 million. In 1982, Grenada exported 2,040 metric tons of nutmeg. Its biggest customers were West Germany (355 metric tons), Britain (332 metric tons), the Soviet Union (299 metric tons), the United States (201 metric tons), Belgium (131 metric tons), the Netherlands (130 metric tons), and Canada (106 metric tons). Other countries imported 486 metric tons of Grenadian nutmeg. In 1982, world production of nutmeg was 10,263 metric tons. Indonesia produced 7,723 of these tons; Grenada, 2,040; other countries, 500. Translate these data into several visuals.

2. Based on data supplied by *Sales and Marketing Management Magazine,* the average cost of daily lodging and meals in 1983 was as follows in the cities indicated: Cleveland, $66.10; New York, $116.60; Orlando, $71.20; San Francisco, $114.65; St. Louis, $75.25; Chicago, $106.75; Kansas City, $75.55; Washington, D. C., $104.00; Los Angeles, $100.25; Toronto, $99.75; Miami, $95.05; Dallas, $76.25; Baltimore, $76.45; New Orleans, $92.75; Boston, $88.40; Atlanta, $84.85; Denver, $83.65; Minneapolis, $84.25; Pittsburgh, $77.30; San Diego, $78.90; Philadelphia, $79.55; Detroit, $81.45; Houston, $83.60. Translate these data into two different visuals.

3. Alumni/ae giving from 1971 to 1980 was as follows (dollars are in thousands): in 1971, 1,263 gave 77,209; in 1972, 1,358 gave 111,014; in 1973, 1,690 gave 131,092; in 1974, 1,937 gave 150,917; in 1975, 2,076 gave 167,692; in 1976, 2,370 gave 195,467; in 1977, 2,289 gave 226,534; in 1978, 2,532 gave 286,965; in 1979, 2,674 gave 313,831; in 1980, 2,927 gave 385,534. Translate these data into a visual.

4. A private high school reported that in 1983 income from tuition came to $1,724,000. Income from fund raising came to $441,000, of which only $95,000 was in alumni giving. Alumni giving for that year came 53.2 percent from the graduating classes

---

[5] P. C. Whalley and R. W. Fleming, "An Experiment with a Simple Recorder of Reading Behaviour," *Programmed Learning and Educational Technology,* 12 (1975), 120–123.

of '69–'73, 4.9 percent from the graduating classes of '79–'83, 18.6 percent from the graduating classes of '64–'68, and 23.3 percent from the graduating classes of '74–'78. Translate these data into a visual.

5. Statistics supplied by the New York City Department of Corrections, the New York City Police Department, and the New York State Division of Substance Abuse Services revealed the following data on the use of drugs in the metropolitan New York region: Police estimate that 24 percent of all slayings committed in the city in 1981 were drug related. The typical heroin addict in the city commits 209 crimes a year, excluding possession and sale of drugs. These crimes include 91 thefts, 34 burglaries, 12 robberies, and 72 other crimes including forgery, fraud, prostitution, and assault. The number of heroin addicts in city jails has increased steadily. There were 7,679 in 1980, 8,745 in 1981, 12,356 in 1982, and 13,046 in 1983. There also were 19,000 heavy heroin users in the city, 32,000 on Long Island and in the northern suburbs, and 20,000 in northern New Jersey. Translate these data into one or more visuals.

6. The Chicago Department of Transportation reports that 19.4 million passengers departed by plane from the city's O'Hare Airport: 4.34 million on American, 1.31 million on Delta, .51 million on Eastern, 1.11 million on Northwest Orient, .33 million on Ozark, .31 million on Piedmont, 1.26 million on Republic, 1.34 million on TWA, 6.94 million on United, .28 million on US Air, and 1.67 on other carriers. Translate these data into a visual.

7. Draw up an organizational chart for your major department at your college or university. On the chart, indicate both positions and the names of those filling positions.

8. Obtain enrollment statistics from your school and then translate them into visuals. You may want to break down total enrollment into any of the following categories: enrollment by sex, by age, by school or college, by year, by full- or part-time status, by religion, by year in school, by major, by means of financing education, by race, by home address, by type of high school attended, by age, by participation in extracurricular activities, by utilization of various school services.

9. Study one or several corporate annual reports (available in many business collections in college and university libraries or by writing to the company itself) to see how well the report makes use of visuals. Then write an evaluative report on the use of such aids following directions supplied by your instructor.

CHAPTER 10

Summaries and Abstracts

In this chapter, we discuss two additional writing products, the *summary* and the *abstract*. Both can play a significant role in helping you communicate with your readers. Both provide your readers with needed shortcuts. Both show you how to save your readers time when you want them to read what you have written.

The summary and the abstract are not substitutes for longer documents. They are simply shortcuts enabling readers to get at the information they need as quickly as possible.

In Part I, we pointed out that all readers share a common characteristic: *All readers are busy.* We also mentioned the results of a study of the reading habits of busy managers at Westinghouse; these managers relied on summaries as shortcuts when they had to read long technical documents.[1]

Those managers did not disregard the longer technical document. They needed that document further down the line when specifics were required or when trouble developed. But, because they were busy, they didn't want

---

[1] See Richard W. Dodge's often-reprinted article, "What to Report," *Westinghouse Engineer,* 22 (July–September, 1962), 108–111. We reprint this article ourselves as part of an exercise at the end of this chapter.

to read twenty or more pages just to find out the main points in the longer document.

On the job, managers and support staff alike have to read memo after memo, letter after letter, and report after report. The further a business person rises in the corporate structure, the more he or she may have to read—and the less time he or she will have to do all that reading. Busy readers then have to depend on summaries.

Summaries and Abstracts: An Initial Distinction

Summaries and abstracts are simply shorter versions of longer documents. These longer documents can be letters, memos, reports, journal articles, manuals, even books.

Business, technical, and professional writing textbooks often distinguish among a variety of writing products when talking about summaries and abstracts. The products include:

Briefs
Descriptive abstracts
Digests
Executive summaries
Informative abstracts
Reviews
Synopses

For the sake of clarity, we will simply use the terms "summary" and "abstract," and we will distinguish between the two products on the basis of *purpose* and *audience*.

A *summary* (sometimes called an *informative abstract*) is a specific, detailed overview of a longer document. A summary includes information about the contents of the longer document and specifics about any conclusions or recommendations the longer document contains.

An *abstract* (sometimes called a *descriptive abstract*) is a general, short overview of a longer document. An abstract simply describes what a longer document discusses. It is concerned with content, not with conclusions or recommendations contained in the original document. An abstract simply expands briefly upon the table of contents of the longer document.

Summaries and abstracts differ then in their purpose. A summary lets a busy reader know what a longer document says. An abstract lets a busy reader know whether or not to spend the time reading the longer document.

Summaries and abstracts—at least in terms of nomenclature—may also differ in their audience. In our own work as writing consultants to various businesses and industries, we have found that readers who consider themselves business people are more familiar with the term *summary*. Readers who consider themselves scientific or technical personnel are more familiar with the term *abstract.*

Throughout this chapter, we will, however, reserve the term *summary* for the more complete overview of the longer document.

Summaries and the PAFEO Process

Purpose and Audience

As with other writing products, summaries have definite purposes. We have already distinguished between summaries and abstracts in terms of how much information they present to readers and of how readers use them to save time. The summary provides a detailed overview for the reader who is too busy to read the longer document. The abstract provides a more general overview for the reader who is trying to decide if he or she needs to read the longer document.

Purpose and audience go hand in hand in another way when you sit down to write a summary. The summary can help you communicate more efficiently and more effectively with your audience in two ways.

First, the summary makes the longer document more immediately accessible to your primary audience. Second, the summary can broaden the audience for a longer document, thereby making information more immediately accessible to a secondary audience. Such an audience will be as busy as your primary audience, but will often lack the technical expertise or the necessary background to read the longer document. Your boss might be your primary audience, for instance; your boss's boss might be a potential secondary audience. In other cases, your boss may be your primary audience while your secondary audience might be co-workers or people outside your organization or company.

As we indicated at the beginning of this chapter, writing summaries and abstracts acknowledges an important fact about your reader. Your reader is busy. The higher your readers go in companies and organizations, the more they need to be able to think, visualize, and plan for several departments or divisions. Such readers need information that is accessible. They are too busy to wade through documents trying to find what they need. They need information at their fingertips.

People further down the organizational ladder or structure have similar needs. They too are busy. Remember, all those Westinghouse engineers wanted information in brief, meaningful, and concise terms at the beginning of a document. Every manager interviewed read the abstract; a bare majority read the introduction, background, conclusions, and recommendations. Only a few read the body of the report or an appendix.

Managers differ in their expertise, not in their reading habits and their needs. A well-written summary is, first of all, a self-contained synopsis that retains only the basic ideas of the longer document. It presents the bottom line while eliminating technical and supporting details and explanations from the longer document. It is, therefore, as we have indicated above, of value to both primary and secondary audiences.

A report, for example, on flex-time, a system of straddled working hours allowing employees some freedom in scheduling their daily work, might be loaded down with technical discussions that would make sense only to the comptroller or the vice-president of personnel. But such a report would have an impact on and be of interest to all employees, many of whom would get bogged down or be lost in technical discussions. A carefully written summary can open up the main points of the report on flex-time to everyone affected by that report's discussions and recommendations.

Format, Evidence, and Organization

Summaries are self-contained. They are not a variation on the concluding paragraph you may be familiar with from other kinds of writing. A summary appears on a separate sheet—it can even run longer than one page—at the beginning or the ending of the longer document, whichever company practice dictates.

In either case, the summary must be clearly labeled and immediately accessible to your readers. Short sentences and short paragraphs are just as important to summaries as they are to longer documents. Headings and the other formatting devices we discussed in Part I further speed your readers on their way through the information they need.

In terms of evidence, a summary cannot contain anything not also contained in the longer document. Rather, a summary simply contains the significant points contained in the longer document. The summary presents "the bottom line."

In writing a summary, you can leave out:

background discussions
personal comments
digressions
conjectures
introductions
explanations
examples, especially lengthy ones
visuals
definitions, especially long or complicated ones
data supported by assertions rather than evidence.[2]

Part of your purpose in providing a summary is to simplify the presentation of information to your readers. Obviously, if all your readers are experts,

[2] We have adapted this list from one provided by John M. Lannon in *Technical Writing*, 3rd ed. (Boston: Little, Brown, 1985), p. 96.

you will not need to do much simplifying for them—although expert readers need summaries as much as nonexpert readers do. When writing a summary for a nonexpert audience or for an audience with mixed levels of expertise, it is safer to simplify rather than to run the risk of losing a segment of your audience.

How much evidence you include in a summary depends in large part on the earlier distinction we made between a summary and an abstract. The abstract gives little more than an annotated table of contents allowing the reader to know if the longer document must be read. The summary presents more in the way of evidence—contents in brief, as well as conclusions and recommendations—saving the reader time when he or she cannot tackle the entire longer document.

Here's an abstract of a 91-page government report on capital punishment. The abstract is designed to let the reader—in this case, perhaps a researcher—determine whether the report contains any information of relevance to his or her research.

CAPITAL PUNISHMENT 1979, BUREAU OF JUST. STATIST. U S DEPT. OF JUST. (WASHINGTON, DC) 1979 (91 PAGES).

This report provides information on prisoners under sentence of death in calendar year 1979, on executions carried out during the 1930–79 period, and on recent trends in the evolution of capital punishment laws. Inmates under sentence of death are differentiated by age, sex, race, marital status at time of imprisonment, and level of education. A count of Hispanic prisoners is also provided. Criminal justice matters treated include legal status at time of arrest, prior felony convictions, offense, time spent on death row, and, for prisoners removed from death row during the year, both the method of removal and status at year end. Appendix I contains the status of death penalty statutes, by jurisdiction: December 31, 1979. The statistical data tables on which the report is based are displayed in Appendix II. Appendix III consists of the basic questionnaire used to elicit data from correctional authorities. A description of the data collection procedures and an account of changes made in the questionnaire in the last several years are presented in Appendix IV, Methodology.[3]

If your reader needs more than a simple overview, you should provide your reader with a summary. A summary will be more complete than an abstract in terms both of what evidence and how much evidence it presents. The summary may be all your busy reader has time to read.

Here is a summary of an article[4] on job prospects for college graduates during the 1980s.

College graduates entering the labor force during the 1980s are expected to encounter market conditions very similar to those traced by entrants in the 1970s. About 15 million college graduates are projected to enter the labor force—about

---

[3] *Criminology and Penology Abstracts,* 22 (March/April, 1982), 170.
[4] J. Sargent, "The Job Outlook for College Graduates during the 1980s," *Occupational Outlook Quarterly,* 26 (Summer 1982), 2–7.

60 percent are expected to be new graduates. Most of the remainder are expected to be reentrants—college-educated workers who left the labor force to raise a family, to pursue graduate education, or for other reasons. Depending on the amount of economic growth realized by the economy as a whole and employment growth in college-graduate–dominated occupations in particular, between 12 and 13 million graduates are projected to be required during the 1980s. About 67 percent of the graduates are expected to be required in professional and technical occupations and 28 percent in managerial, administrative, and sales occupations. A surplus of between 2 and 3 million college graduates is expected to enter the labor force during the 1980s.[5]

Length is not an issue in writing a summary. Completeness is, and completeness can go hand in hand with considerations of format. The following summary is complete. It runs a little more than 200 words, but it is poorly formatted. The first paragraph looks more formidable than it really is. The second paragraph is only slightly better.

The purpose of this summary is to save the reader time. These three paragraphs will certainly take less time to read than the entire report. But there is nothing about the format—or the organization—of the summary that helps the reader through what is a piece of writing specifically intended to save a busy reader time.

SUMMARY

As requested, we have reviewed the pricing structure for data processing services provided by Paulcomp, Inc. to their parent company, Paulis Publishing, Inc. and other subsidiaries of Paulis Publishing. In performing this review, we have uncovered a number of problem areas not all directly related to the pricing structure. Although pricing is the most relevant issue to be considered by Paulcomp at this time, there are other procedural improvements that must be addressed concurrently. This is necessary to improve service and profitability of the current clients immediately, and to support Paulcomp's long-range business goal of marketing their services externally (outside the confines of Paulis Publishing, Inc.).

The problem areas appear to be billing, collection, cost accounting, and quality control. We recommend that management take immediate action directed towards achieving increased efficiency and effectiveness in these areas, particularly billing and collection. These actions should be initiated simultaneously with development of a new pricing algorithm so that increased user confidence, which should result from these improvements, will help ease any user resistance to implementation of a new pricing structure.

Benefits to be accrued by Paulcomp will be improved services to its customers, consistency in service and pricing, elimination of billing errors, and increased revenue through equitable billing practices.

Here's a second version of the summary. This version uses format and organization to present evidence the reader needs and to accomplish the

---

[5] This summary appears in *Human Resource Abstracts,* 18 (September, 1983), 301.

writer's purpose. The report itself is divided into sections, each clearly labeled. The headings from these sections of the report have been retained in the summary, whose presentation of evidence is enhanced further for the busy reader by the use of shorter paragraphs and enumerated vertical lists.

Summary

PURPOSE AND SCOPE
As requested, we have reviewed the pricing structure for data processing services provided by Paulcomp, Inc. to the parent company, Paulis Publishing, Inc. and other subsidiaries of Paulis Publishing.

FINDINGS
In addition to discovering serious problems in pricing policy, our review also brought to light several additional problems which must be addressed concurrently.

These problems are found in:

- billing
- collection
- cost accounting
- quality control

CONCLUSIONS
Improving policies and procedures in these four areas is a must. Not only will such improvement result in better service to current clients, but it will also support Paulcomp's long-range business goal of marketing their services externally to clients outside the corporation.

In addition to bettering service, the needed improvements will benefit Paulcomp by:

1. providing consistency in service and pricing
2. eliminating billing errors
3. increasing revenue through equitable billing practices.

RECOMMENDATIONS
We recommend that management take immediate action to achieve increased efficiency and effectiveness in billing, collection, cost accounting, and quality control.

These improvements should go hand-in-hand with the development of a new pricing algorithm. The increased confidence of users which will result from the improvements will also reduce resistance to needed changes in the pricing structure.

In Chapter 12, we will return to this summary. The report upon which it is based is our sample report in that chapter.

A Procedure for Writing Summaries

With some practice, summaries are easy to write. Here is an easy-to-follow procedure for writing summaries.

1. Read the entire document to develop a sense of both what it says and how it says it. (As the writer of the longer document, you may have become so involved with the writing that you can no longer remain objective.

2. Read the document a second time, underlining the main points or eliminating the secondary and supporting points. (In a well-written document, these main points may simply be topic or thesis sentences.)
   a. Concentrate on basic ideas only.
   b. Eliminate supporting facts, examples, or data.
   c. Eliminate illustrations and bibliographic references.
   d. Combine similar ideas into more direct single statements.
   e. Avoid technical terminology.
   f. Condense information as concisely as possible.

3. Rewrite what's left from the original, whether or not you authored the original document.
   a. Begin with a clear statement of the longer document's main idea.
   b. Supply any necessary transition between the points in your abstract.
   c. When necessary, break your abstract up into logical paragraphs that parallel the organizational pattern of the longer documents.
   d. Don't hesitate to use any of the formatting tricks we introduced earlier in this textbook.

When followed correctly, this procedure will allow you to condense any longer document systematically. No document can contain hundreds and hundreds of ideas. Instead, it will contain a limited number of ideas that you discuss, develop, argue for or against, and provide examples of for your reader.

In writing a summary, you will be zeroing in on that limited number of ideas. If your readers need more information or have more time, they can turn to the longer document after reading the summary.

An Application of This Procedure

Any document will contain a limited number of ideas. What follows is an essay from *Time*.[6] The essay, a variation on an op-ed piece, shares an important feature with any business or technical document. In its sixteen paragraphs, it contains only a few basic, important ideas. The essay isn't badly written; it isn't padded. It simply requires sixteen paragraphs to develop the points it wants to make.

Much of the prose in these paragraphs is given over to supporting informa-

---

[6] Reprinted by permission, from *Time*, 113 (May 14, 1979), 54, 59.

tion—examples, testimonials and quotations, repetitions for the sake of emphasis—that a busy reader wouldn't have time to read. What a busy reader might have time to read, though, is a summary. First, the essay itself.

A New Distrust of the Experts

"Whenever the people are well informed, they can be trusted with their own government." Thomas Jefferson's axiom remains an indispensable premise of democracy. Yet the possibility of a sage and knowing public seems to be growing ever more elusive. Since the rise of science and technology as the commanding force in both government and social change, it has become harder and harder for most Americans to become really well informed on the problems they face as individuals or citizens. Such a trend is bound to raise questions about the future of popular rule.

Nowadays the very vocabulary of public discourse can be bewildering. Even to be half informed, the American-on-the-street must grasp terms like deoxyribonucleic acid, fantastic prospects like genetic engineering, and bizarre phenomena like nuclear meltdown. The technical face of things has driven some people into a bored sort of cop-out—"science anxiety," it is called by Physics Professor Jeffry Mallow of Loyola University in Chicago. The predicament has made most Americans hostage to the superior knowledge of the expert: the scientist, the technician, the engineer, the specialist.

Society has grown so complicated that there is renewed interest in the possibility of a "science court" that might deal impartially with arcane controversy. It has grown so technical that some lawyers wonder whether ordinary electors can still adequately function as jurors. Says Attorney Gary Ahrens, a professor at the University of Iowa: "Practically nothing is commonsensical any more." Surely the spectacle of the public making decisions in semi-darkness is an affront to common sense.

Dependency on the experts seemed tenable in the more innocent era when science was viewed as a virtually infallible cornucopia of social goodies. Americans long clung to Virgil's ancient advice: "Believe an expert." Today, however, Americans are no longer willing to acquiesce gratefully in either the discoveries of science or their application. The citizen has rediscovered that the best of experts will now and then launch an unsinkable *Titanic*.

The public has needed no expertise to read about DDT, thalidomide and cyclamates, nor to learn that the DES that seemed a nifty preventive of miscarriage in the 1950s was being linked to cancer a generation later. The citizen's problem, at bottom, is how to assess the things that so often come forth in the beguiling guise of blessings. What to believe? Whom to trust? This is a recipe for public frustration.

The shadow of science falls across decisions common to daily existence. Is this medication safe? Is forgoing sugar worth the hazards of saccharin? Are the conveniences of the Pill worth raising the risk of circulatory disease? The uncertain answers come from product analysts, dietitians, pharmacists, lawyers, physicians. American society, as Federal Trade Commission Chairman Michael Pertschuk puts it, has become "dominated by professionals who call us 'clients' and tell us of our 'needs.'"

The biggest problem, however, is that the faith of the American people in the experts has been badly shaken. People have learned, for one thing, that certified technical gospel is far from immortal. Medicine changes its mind about tonsillectomies that used to be routinely performed. Those dazzling phosphate detergents turn out to be anathema to the environment. Scarcely a week goes by without the credibility of one expert or another falling afoul of some spike of fresh news. (Just last week an array of nonprescription sedatives used by millions was linked, through the ingredient methapyrilene, to cancer.) Moreover, experts are constantly challenging experts, debating the benefits and hazards of virtually every technical thrust. Who knows anything for sure? Could supersonic aircraft truly damage the ozone? The technical sages disagree.

Thus the problems that the individual copes with as a private person are knotty enough; public issues have grown immeasurably more complex. Government has long since subsumed science and technology into its realm, both as the fountainhead of its projects and as an object of its regulation. The calculations that measure national military strength are as impenetrable to the civilian-on-the-street as the formulas of the ancient alchemists. The surreal arithmetic of SALT might as well be the music of the spheres, for all the help it gives ordinary folks trying to get a clear picture of the country's real and relative strengths. The nervous strategist is not the only one to covet verification; the common citizen could also use some.

Then, too, much information crucial to the personal and social decisions of citizens is methodically hidden or withheld. The scientific world has always tended to hoard lore on work in progress, and the Government's customary secrecy in military matters, intelligence and foreign affairs has spread to many parts of the bureaucratic and corporate spheres. The clandestine spirit that properly cloaked the devising of atomic weapons inevitably carried over to veil the development of nuclear power for civilian purposes.

The result of secrecy compounded by confusion and some startling ignorance was dramatized by the Three Mile Island nuclear power plant crisis. While the event made plain that Government and corporate experts had not quite leveled with the public about the hazards of nuclear power, it also proved, frighteningly enough, that the experts sometimes did not tell the whole story simply because they did not know it. Joseph M. Hendrie, chairman of the Nuclear Regulatory Commission, said of himself and other officials, as they tried to cope with an incipient meltdown: "We are operating . . . like a couple of blind men staggering around making decisions."

Intentional deception sometimes leaves the citizenry in a plight as awkward as Hendrie's. Last month a former ranking employee charged that the Hooker Chemicals and Plastics Corp. of Niagara Falls, N. Y., had kept workers in the dark about the hazards of toxic chemicals they dealt with. Federal atomic authorities, it was disclosed last month, were encouraged by President Dwight Eisenhower to confuse the public about the risks of radiation fallout during the atomic bomb tests in Nevada in the 1950s; Government officials refused to warn inhabitants of nearby regions that they were absorbing possibly lethal doses of radiation.

The citizenry's essential interest is not in knowledge *per se* but the social uses to which it is put. What is often kept from the citizen, in the form of knowledge, is social and political power. When demonstrations and controversies break out over seemingly esoteric technical questions, the underlying question, as Cornell University's Dorothy Nelkin puts it in a paper on "Science as a Source of Political Conflict," is always the same: "Who should control crucial policy choices?" Such choices, she adds, tend to stay in the hands of those who control "the context of facts and values in which policies are shaped."

On its face, the situation may help explain the mood of public disenchantment that has persisted long after the events—Viet Nam and Watergate—that were supposed to have caused it. Surely neither of those national traumas caused the drop of popular confidence in almost all key U. S. institutions that Pollster Louis Harris recently recorded. It also seems doubtful that either deprived the Administration's energy crusade of both popular support and belief. Could it be that many citizens simply feel foreclosed not only from knowledge but also from the power that knowledge would give them?

The public itself, it must be admitted, bears a fair share of responsibility for its dilemma. It has usually welcomed the advances and conveniences—swift travel, cheap energy, life-prolonging medication, magical cosmetics—and left itself no choice but to live with the inherent risks it does not so cheerfully accept. A completely risk-free society would be a dead society. In today's increasingly risk-shy atmosphere, the public may tend to exaggerate some of the dangers at hand. Indeed, it may be swinging from too much awe of the "miracles" of science and technology to excessive skepticism about them. In reality, the public has always wanted to lean on the experts—until they have failed, or seemed to.

It is fair to suppose that even if the public had access to all knowledge about everything, there would still be a good deal of befuddlement and groping. Not many have the ability, energy and will to bone up on every issue. If it is reasonable for Americans to demand more candor, prudence—and humility—from the experts, it is also reasonable that the citizenry demand of itself ever greater diligence in using all available information, including journalism's increasingly technical harvest.

Plainly the citizen's plight is not subject to quickie remedy. Yet any solution would have to entail a shift in the relationship between the priests of knowledge and the lay public. The expert will have to play a more conscious role as citizen, just as the ordinary American will have to become ever more a student of technical lore. The learned elite will doubtless remain indispensable. Still, the fact that they are exalted over the public should not mean that they are excused from responsibility to it—not unless the Jeffersonian notion of popular self-rule is to be lost be default.   —*Frank Trippett*

Here's the same essay again. Only main points are underlined, with the result that the text of the essay has been shortened immediately by over 50 percent.

A New Distrust of the Experts

"Whenever the people are well informed, they can be trusted with their own government." Thomas Jefferson's axiom remains an indispensable premise of democracy. Yet the possibility of a sage and knowing public seems to be growing ever more elusive. <u>Since the rise of science and technology as the commanding</u>

force in both government and social change, it has become harder and harder for most Americans to become really well informed on the problems they face as individuals or citizens. Such a trend is bound to raise questions about the future of popular rule.

Nowadays the very vocabulary of public discourse can be bewildering. Even to be half informed, the American-on-the-street must grasp terms like deoxyribonucleic acid, fantastic prospects like genetic engineering, and bizarre phenomena like nuclear meltdown. The technical face of things has driven some people into a bored sort of cop-out—"science anxiety," it is called by Physics Professor Jeffry Mallow of Loyola University in Chicago. The predicament has made most Americans hostage to the superior knowledge of the expert: the scientist, the technician, the engineer, the specialist.

Society has grown so complicated that there is renewed interest in the possibility of a "science court" that might deal impartially with arcane controversy. It has grown so technical that some lawyers wonder whether ordinary electors can still adequately function as jurors. Says Attorney Gary Ahrens, a professor at the University of Iowa: "Practically nothing is commonsensical any more." Surely the spectacle of the public making decisions in semi-darkness is an affront to common sense.

Dependency on the experts seemed tenable in the more innocent era when science was viewed as a virtually infallible cornucopia of social goodies. Americans long clung to Virgil's ancient advice: "Believe an expert." Today, however, Americans are no longer willing to acquiesce gratefully in either the discoveries of science or their application. The citizen has rediscovered that the best of experts will now and then launch an unsinkable *Titanic*.

The public has needed no expertise to read about DDT, thalidomide and cyclamates, nor to learn that the DES that seemed a nifty preventive of miscarriage in the 1950s was being linked to cancer a generation later. The citizen's problem, at bottom, is how to assess the things that so often come forth in the beguiling guise of blessings. What to believe? Whom to trust? This is a recipe for public frustration.

The shadow of science falls across decisions common to daily existence. Is this medication safe? Is forgoing sugar worth the hazards of saccharin? Are the conveniences of the Pill worth raising the risk of circulatory disease? The uncertain answers come from product analysts, dietitians, pharmacists, lawyers, physicians. American society, as Federal Trade Commission Chairman Michael Pertschuk puts it, has become "dominated by professionals who call us 'clients' and tell us of our 'needs.'"

The biggest problem, however, is that the faith of the American people in the experts has been badly shaken. People have learned, for one thing, that certified technical gospel is far from immortal. Medicine changes its mind about tonsillectomies that used to be routinely performed. Those dazzling phosphate detergents turn out to be anathema to the environment. Scarcely a week goes by without the credibility of one expert or another falling afoul of some spike of fresh news. (Just last week an array of nonprescription sedatives used by millions was linked, through the ingredient methapyrilene, to cancer.) Moreover, experts are constantly challenging experts, debating the benefits and hazards of virtually every technical thrust. Who knows anything for sure? Could supersonic aircraft truly damage the ozone? The technical sages disagree.

Thus the problems that the individual copes with as a private person are knotty enough; public issues have grown immeasurably more complex. Government has long since subsumed science and technology into its realm, both as the fountainhead of its projects and as an object of its regulation. The calculations that measure national military strength are as impenetrable to the civilian-on-the-street as the formulas of the ancient alchemists. The surreal arithmetic of SALT might as well be the music of the spheres, for all the help it gives ordinary folks trying to get a clear picture of the country's real and relative strengths. The nervous strategist is not the only one to covet verification; the common citizen could also use some.

Then, too, much information crucial to the personal and social decisions of citizens is methodically hidden or withheld. The scientific world has always tended to hoard lore on work in progress, and the Government's customary secrecy in military matters, intelligence and foreign affairs has spread to many parts of the bureaucratic and corporate spheres. The clandestine spirit that properly cloaked the devising of atomic weapons inevitably carried over to veil the development of nuclear power for civilian purposes.

The result of secrecy compounded by confusion and some startling ignorance was dramatized by the Three Mile Island nuclear power plant crisis. While the event made plain that Government and corporate experts had not quite leveled with the public about the hazards of nuclear power, it also proved, frighteningly enough, that the experts sometimes did not tell the whole story simply because they did not know it. Joseph M. Hendrie, chairman of the Nuclear Regulatory Commission, said of himself and other officials, as they tried to cope with an incipient meltdown: "We are operating . . . like a couple of blind men staggering around making decisions."

Intentional deception sometimes leaves the citizenry in a plight as awkward as Hendrie's. Last month a former ranking employee charged that the Hooker Chemicals and Plastics Corp. of Niagara Falls, N. Y., had kept workers in the dark about the hazards of toxic chemicals they dealt with. Federal atomic authorities, it was disclosed last month, were encouraged by President Dwight Eisenhower to confuse the public about the risks of radiation fallout during the atomic bomb tests in Nevada in the 1950s; Government officials refused to warn inhabitants of nearby regions that they were absorbing possibly lethal doses of radiation.

The citizenry's essential interest is not in knowledge *per se* but the social uses to which it is put. What is often kept from the citizen, in the form of knowledge, is social and political power. When demonstrations and controversies break out over seemingly esoteric technical questions, the underlying question, as Cornell University's Dorothy Nelkin puts it in a paper on "Science as a Source of Political Conflict," is always the same: "Who should control crucial policy choices?" Such choices, she adds, tend to stay in the hands of those who control "the context of facts and values in which policies are shaped."

On its face, the situation may help explain the mood of public disenchantment that has persisted long after the events—Viet Nam and Watergate—that were supposed to have caused it. Surely neither of those national traumas caused the drop of popular confidence in almost all key U. S. institutions that Pollster Louis Harris recently recorded. It also seems doubtful that either deprived the Administration's energy crusade of both popular support and belief. Could it be that many citizens simply feel foreclosed not only from knowledge but also from the power that knowledge would give them?

The public itself, it must be admitted, bears a fair share of responsibility for its dilemma. It has usually welcomed the advances and conveniences—swift travel, cheap energy, life-prolonging medication, magical cosmetics—and left itself no choice but to live with the inherent risks it does not so cheerfully accept. A completely risk-free society would be a dead society. In today's increasingly risk-shy atmosphere, the public may tend to exaggerate some of the dangers at hand. Indeed, it may be swinging from too much awe of the "miracles" of science and technology to excessive skepticism about them. In reality, the public has always wanted to lean on the experts—until they have failed or seemed to.

It is fair to suppose that even if the public had access to all knowledge about everything, there would still be a good deal of befuddlement and groping. Not many have the ability, energy and will to bone up on every issue. If it is reasonable for Americans to demand more candor, prudence—and humility—from the experts, it is also reasonable that the citizenry demand of itself ever greater diligence in using all available information, including journalism's increasingly technical harvest.

Plainly the citizen's plight is not subject to quickie remedy. Yet any solution would have to entail a shift in the relationship between the priests of knowledge and the lay public. The expert will have to play a more conscious role as citizen, just as the ordinary American will have to become ever more a student of technical lore. The learned elite will doubtless remain indispensable. Still, the fact that they are exalted over the public should not mean that they are excused from responsibility to it—not unless the Jeffersonian notion of popular self-rule is to be lost by default.   —*Frank Trippett*

But you can do more than reduce the text of this document by 50 percent. Look at the underlined sections of the essay again. As you do, apply the subsequent steps in the procedure we provided for writing summaries. Concentrate on basic ideas. Eliminate supporting points. Combine similar points into more direct simple statements.

Here is a possible five-sentence summary of the *Time* essay:

> The language of technology and technology itself often leave the public in the dark and thereby threaten democracy. The public is in the dark because the experts are sometimes there too. Government tends to complicate the problem further through unnecessary secrecy and downright lies. To be responsible members of society, the public needs to be informed. For such information, the public and the experts must work together and lessen the threat to democracy that public ignorance entails.

As we said, this is only a possible version of a summary. You might come up with different wording, or one fewer or one more sentence. But your version of the summary should highlight the same main points as the one printed above.

Here's a diagram showing how we eventually moved from the sixteen paragraphs of the original *Time* essay to our five-sentence summary:

	TIME ESSAY	SUMMARY
Paragraphs	1 2 3	are summarized in Sentence 1.
	4 5 6 7	are summarized in Sentence 2.
	8 9 10 11	are summarized in Sentence 3.
	12 13	are summarized in Sentence 4.
	14 15 16	are summarized in Sentence 5.

Summaries Help Writers, Too

So far in this chapter, we have discussed the advantages that summaries have for readers. They save time for primary audiences, and they make documents otherwise inaccessible accessible to secondary audiences. Summaries can also benefit writers of longer documents.

As you may have noticed, our procedure for writing summaries essentially asks you to outline the longer document *after* you have written it. Outlining is a part of the writing process that you may already use when you are getting ready to write. Such an outline can help you organize what you want to say before attempting a first draft.

If, however, you follow the procedure for writing summaries we provide and you cannot produce a logical outline of a longer document you have written, you may have just told yourself that your longer document really won't make much sense to your reader or that your longer document is poorly organized.

We are all always our own worst editors. The more time we spend on a writing project, the less objectivity we can develop about that project. By outlining in reverse to produce a summary, you can provide yourself with one last objective review of your longer document to make sure that it says what you intend it to say clearly and logically. In short, by outlining in reverse, you can make sure that your format and organization support your evidence, and that these three combine to achieve your purpose in writing to your reader. Writing a summary can be a simple extension of the PAFEO process.

Abstracts

Depending on your major or your areas of expertise, you may already be familiar with abstracts through the work of professional abstracting services. Such services employ staff to read all the literature in a given field and then to describe the contents of each study, paper, review, journal article, book, or other publication. These abstracts are invaluable to readers who wouldn't have time to read several hundred documents to determine if they contained any relevant information. Such readers can instead simply scan a set of abstracts that will let them know what to read in full and what to ignore.

Another source for such abstracts may be *annotated bibliographies.* These bibliographies list publications in a field and then describe their contents in brief. Such bibliographies may be published annually by professional organizations, or they may appear in article or book form as separate publications independent of any professional organization.

Here is an abstract from an annotated bibliography of books published on business and technical writing. The abstract is of Robert Gunning's *The Technique of Clear Writing,* rev. ed. (New York: McGraw-Hill, 1968).

> This book, originally published in 1952, is divided into three parts. Part I, "What Your Reader Wants," discusses in three sections the danger of clouded or "foggy" writing, analyzing the reader, and various readability formulas, including the author's own Fog Index. Part II, "Ten Principles of Clear Writing," covers style and reader analysis. Part III, "Causes and Cures," covers newspaper, business, legal, and technical writing. The book concludes with three appendixes: a readability analysis of the Gettysburg Address, the Edgar Dale List of 3,000 Familiar Words, and a list of long words with short-word alternatives.[7]

Abstracts and the Literature Search

If you are still in school, you may be assigned to write an abstract of a document written by someone else. Such an assignment might be part of a review of the literature in a field. (On the job, you are more likely to write summaries. When you have to write an abstract, you may be abstracting a document you or someone else wrote.)

Writing abstracts is fairly simple. Remember that your reader wants nothing more than a simple expanded table of contents. Your reader is not using an abstract for the same purpose as he or she would use a summary. Your

---

[7] Gerald J. Alred, Diana C. Reep, and Mohan R. Limaye, *Business and Technical Writing, An Annotated Bibliography of Books, 1880–1980* (Metuchen, N J: Scarecrow Press, 1981), p. 88.

reader is still busy, but he or she wants simply to know whether to read or to ignore the longer document.

Here is a list of regularly published abstracts compiled by professional abstracting societies or professional organizations in the fields of business, science, and technology. We will refer to this list again in the end-of-chapter exercises.

Abstracts of Crime and Juvenile Delinquency
Abstracts of Health Care Management Studies
Abstracts of Hospital Management Studies
Air Pollution Abstracts
Biological Abstracts
Ceramic Abstracts
Chemical Abstracts
Child Development Abstracts and Bibliography
College Student Personnel Abstracts
Communication Abstracts
Computer Abstracts
Criminal Justice Abstracts
Criminology and Penology Abstracts
Dissertation Abstracts International
Employment Relations Abstracts
Energy Abstracts for Policy Analysis
Energy Information Abstracts
Energy Research Abstracts
Food Service and Technology Abstracts
Historical Abstracts
Human Resources Abstracts
Information Science Abstracts
International Abstracts in Operations Research
International Aerospace Abstracts
International Political Science Abstracts
Journal of Economic Abstracts
Library Science Abstracts
Medical Electronics and Communications Abstracts
Metals Abstracts
Nuclear Science Abstracts
Nutrition Abstracts and Reviews
Personnel Management Abstracts
Physics Abstracts
Psychological Abstracts
Science Abstracts
Social Work Research and Abstracts
Sociological Abstracts
Statistical Theory and Method Abstracts
Urban Affairs Abstracts
Work Related Abstracts
World Textile Abstracts

Exercises

1. The letter from the president or chair of the board in a corporate annual report often functions as a summary of what follows. Your local or school library will have a collection of these annual reports. Consult the reports from several companies to see how effectively the opening letter functions as a summary of the longer report. If your instructor directs you to do so, prepare a written or oral report commenting on how effective these letters are as summaries.

2. Visit your local or school library to determine which of the published abstracts listed on page 237 your library receives. Make a list of any other abstracts to which your library subscribes. If your instructor directs you to do so, prepare a memo to your class or to the instructor on how extensive your library's holdings are in the area of published abstracts.

3. Here is the article on the reading habits of managers at Westinghouse that we have discussed at several points earlier in this textbook.[8] Read it as a basis for classroom discussion. Then, if your instructor directs you to, write a summary and an abstract of this article.

What to Report

Technical reports *can* be a useful tool for management—not only as a source of general information, but, more importantly, as a valuable aid to decision making. But to be effective for these purposes, a technical report must be geared to the needs of management.

Considerable effort is being spent today to upgrade the effectiveness of the technical report. Much of this is directed toward improving the writing abilities of engineers and scientists; or toward systems of organization, or format, for reports. This effort has had some rewarding effects in producing better written reports.

But one basic factor in achieving better reports seems to have received comparatively little attention. This is the question of audience needs. Or, expressed another way, *"What does management want in reports?"* This is an extremely basic question, and yet it seems to have had less attention than have the mechanics of putting words on paper.

A recent study conducted at Westinghouse sheds considerable light on this subject. While the results are for one company, probably most of them would apply equally to many other companies or organizations.

The study was made by an independent consultant with considerable experience in the field of technical report writing. It consisted of interviews with Westinghouse men at every level of management, carefully selected to present an accurate cross section. The list of questions asked is shown in Table 1. The results were compiled and analyzed and from the report several conclusions are

[8] Reprinted by permission, from Richard W. Dodge, "What to Report," *Westinghouse Engineer*, 22 (July–September, 1962), 108–111.

apparent. In addition, some suggestions for report writers follow as a natural consequence.

WHAT MANAGEMENT LOOKS FOR IN ENGINEERING REPORTS

When a manager reads a report, he looks for pertinent facts and competent opinions that will aid him in decision making. He wants to know right away whether he should read the report, route it, or skip it.

To determine this, he wants answers fast to some or all of the following questions:

- What's the report about and who wrote it?
- What does it contribute?
- What are the conclusions and recommendations?
- What are their importance and significance?
- What's the implication to the Company?
- What actions are suggested? Short range? Long range?
- Why? By whom? When? How?

The manager wants this information in brief, concise, and meaningful terms. He wants it at the beginning of the report and all in one piece.

For example, if a summary is to convey information efficiently, it should contain three kinds of facts:

1. What the report is about;
2. The significance and implications of the work; and
3. The action called for.

TABLE 1.   Questions Asked of Managers

---

1. What types of reports are submitted to you?
2. What do you look for *first* in the reports submitted to you?
3. What do you want from these reports?
4. To what depth do you want to follow any one particular idea?
5. At what level (how technical and how detailed) should the various reports be written?
6. What do you want emphasized in the reports submitted to you? (Facts, interpretations, recommendations, implications, etc.)
7. What types of decisions are you called upon to make or to participate in?
8. What type of information do you need in order to make these decisions?
9. What types of information do you receive that you don't want?
10. What types of information do you want but not receive?
11. How much of a typical or average report you receive is useful?
12. What types of reports do you write?
13. What do you think your boss wants in the reports you send him?
14. What percentage of the reports you receive do you think desirable or useful? (In kind or frequency.)
15. What percentage of the reports you write do you think desirable or useful? (In kind or frequency.)
16. What particular weaknesses have you found in reports?

To give an intelligent idea of what the report is about, first of all the problem must be defined, then the objectives of the project set forth. Next, the reasons for doing the work must be given. Following this should come the conclusions. And finally, the recommendations.

Such summaries are informative and useful, and should be placed at the beginning of the report.

The kind of information a manager wants in a report is determined by his management responsibilities, but how he wants this information presented is determined largely by his reading habits. This study indicates that management report reading habits are surprisingly similar. Every manager interviewed said he read the *summary* or abstract; a bare majority said they read the *introduction* and *background* sections as well as the *conclusions* and *recommendations;* only a few managers read the *body* of the report or the *appendix* material.

The managers who read the *background* section, or the conclusions and recommendations, said they did so ". . . to gain a better perspective of the material being reported and to find an answer to the all-important question: What do we do next?" Those who read the *body* of the report gave one of the following reasons:

1. Especially interested in subject;
2. Deeply involved in the project;
3. Urgency of problem requires it;
4. Skeptical of conclusions drawn.

And those few managers who read the *appendix* material did so to evaluate further the work being reported. To the report writer, this can mean but one thing: If a report is to convey useful information efficiently, the structure must fit the manager's reading habits.

The frequency of reading chart in Fig. 1 suggests how a report should be structured if it is to be useful to management readers.

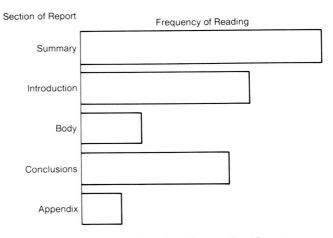

Fig. 1. How Managers Read Reports

SUBJECT MATTER INTEREST

In addition to what facts a manager looks for in a report and how he reads reports, the study indicated that he is interested in five broad technological areas. These are:

1. Technical problems;
2. New projects and products;
3. Experiments and tests;
4. Materials and processes;
5. Field troubles.

Managers want to know a number of things about each of these areas. These are listed in Table 2. Each of the sets of questions can serve as an effective check list for report writers.

TABLE 2.   What Managers Want to Know

Problems	*Tests and Experiments*
What is it?	What tested or investigated?
What undertaken?	Why? How?
Magnitude and importance?	What did it show?
What is being done? By whom?	Better ways?
Approaches used?	Conclusions? Recommendations?
Thorough and complete?	Implications to Company?
Suggested solution? Best? Consider others?	
What now?	*Materials and Processes*
Who does it?	Properties, characteristics, capabilities? Limitations?
Time factors?	Use requirements and environment?
	Areas and scope of application?
New Projects and Products	Cost factors?
Potential?	Availability and sources?
Risks?	What else will do it?
Scope of application?	Problems in using?
Commercial implications?	Significance of application to Company?
Competition?	
Importance to Company?	*Field Troubles and Special Design Problems*
More work to be done? Any problems?	Specific equipment involved?
Required manpower, facilities and equipment?	What trouble developed? Any trouble history?
Relative importance to other projects or products?	How much involved?
Life of project or product line?	Responsibility? Others? Westinghouse?
Effect on Westinghouse technical position?	What is needed?
Priorities required?	Special requirements and environment?
Proposed schedule?	Who does it? Time factors?
Target date?	Most practical solution? Recommended action?
	Suggested product design changes?

In addition to these subjects, a manager must also consider market factors and organization problems. Although these are not the primary concern of the engineer, he should furnish information to management whenever technical aspects provide special evidence or insight into the problem being considered. For example, here are some of the questions about marketing matters a manager will want answered:

- What are the chances for success?
- What are the possible rewards? Monetary? Technological?
- What are the possible risks? Monetary? Technological?
- Can we be competitive? Price? Delivery?
- Is there a market? Must one be developed?
- When will the product be available?

And, here are some of the questions about organization problems a manager must have answered before he can make a decision:

- Is it the type of work Westinghouse should do?
- What changes will be required? Organization? Manpower? Facilities? Equipment?
- Is it an expanding or contracting program?
- What suffers if we concentrate on this?

These are the kinds of questions Westinghouse management wants answered about projects in these five broad technological areas. The report writer should answer them whenever possible.

LEVEL OF PRESENTATION

Trite as it may sound, the technical and detail level at which a report should be written depends upon the reader and his use of the material. Most readers—certainly this is true for management readers—are interested in the significant material and in the general concepts that grow out of detail. Consequently, there is seldom real justification for a highly technical and detailed presentation.

Usually the management reader has an educational and experience background different from that of the writer. *Never* does the management reader have the same knowledge of and familiarity with the specific problem being reported that the writer has.

Therefore, the writer of a report for management should write at a technical level suitable for a reader whose educational and experience background is in a field different from his own. For example, if the report writer is an electrical engineer, he should write his reports for a person educated and trained in a field such as chemical engineering, or mechanical engineering, or metallurgical engineering.

All parts of the report *should preferably* be written on this basis. The highly technical, mathematical, and detailed material—if necessary at all—can and should be placed in the appendix.

MANAGEMENT RESPONSIBILITIES

The information presented thus far is primarily of interest to the report writer. In addition, however, management itself has definite responsibilities in the reporting process. These can be summed up as follows:

1. Define the project and the required reports;
2. Provide proper perspective for the project and the required reporting;
3. See that effective reports are submitted on time; and
4. See that the reports are properly distributed.

An engineering report, like any engineered product, has to be designed to fill a particular need and to achieve a particular purpose within a specific situation. Making sure that the writer knows what his report is to do, how it is to be used, and who is going to use it—all these things are the responsibilities of management. Purpose, use, and reader are the design factors in communications, and unless the writer knows these things, he is in no position to design an effective instrument of communication—be it a report, a memorandum, or what have you.

Four conferences at selected times can help a manager control the writing of those he supervises and will help him get the kind of reports he wants, when he wants them.

Step 1—At the beginning of the project. The purpose of this conference is to define the project, make sure the engineer involved knows what it is he's supposed to do, and specify the required reporting that is going to be expected of him as the project continues. What kind of decisions, for example, hinge upon his report? What is the relation of his work to the decision making process of management? These are the kinds of questions to clear up at this conference.

If the project is an involved one that could easily be misunderstood, the manager may want to check the effectiveness of the conference by asking the engineer to write a memorandum stating in his own words his understanding of the project, how he plans to handle it, and the reporting requirements. This can assure a mutual understanding of the project from the very outset.

Step 2—At the completion of the investigation. When the engineer has finished the project assignment—but before he has reported on it—the manager should have him come in and talk over the results of his work. What did he find out? What conclusions has he reached? What is the main supporting evidence for these conclusions? What recommendations does he make? Should any future action be suggested? What is the value of the work to the Company?

The broader perspective of the manager, plus his extensive knowledge of the Company and its activities, puts him in the position of being able to give the engineer a much better picture of the value and implications of the project.

The mechanism for getting into the report the kind of information needed for decision making is a relatively simple one. As the manager goes over the material with the engineer, he picks out points that need to be emphasized, and those that can be left out. This is a formative process that aids the engineer in the selection of material and evidence to support his material.

Knowing in advance that he has a review session with his supervisor, the chances are the engineer will do some thinking beforehand about the project and the results. Consequently, he will have formed some opinions about the significance of the work, and will, therefore, make a more coherent and intelligent presentation of the project and the results of his investigation.

This review will do something for the manager, too. The material will give him an insight into the value of the work that will enable him to converse intelligently and convincingly about the project to others. Such a preview may, therefore, expedite decisions influencing the project in one way or another.

Step 3—After the report is outlined. The manager should schedule a third conference after the report is outlined. At this session, the manager and the author should review the report outline step-by-step. If the manager is satisfied with the outline, he should tell the author so and tell him to proceed with the report.

If, however, the manager is not satisfied with the outline and believes it will have to be reorganized before the kind of report wanted can be written, he must make this fact known to the author. One way he can do this is to have the author tell him why the outline is structured the way it is. This usually discloses the organizational weakness to the author and consequently he will be the one to suggest a change. This, of course, is the ideal situation. However, if the indirect approach doesn't work, the more direct approach must be used.

Regardless of the method used to develop a satisfactory outline, the thing the manager must keep in mind is this: It's much easier to win the author's consent to structural changes at the outline stage, i.e., *before* the report is written, than afterward. Writing is a personal thing; therefore, when changes are suggested in organization or approach, these are all too frequently considered personal attacks and strained relations result.

Step 4—After the report is written. The fourth interview calls for a review and approval by the manager of the finished report and the preparation of a distribution list. During this review, the manager may find some sections of the report that need changing. While this is to be expected, he should limit the extent of these changes. The true test of any piece of writing is the clarity of the statement. If it's clear and does the job, the manager should leave it alone.

This four-step conference mechanism will save the manager valuable time, and, it will save the engineer valuable time. Also, it will insure meaningful and useful project reports—not an insignificant accomplishment itself. In addition, the process is an educative one. It places in the manager's hands another tool he can use to develop and broaden the viewpoint of the engineer. By eliminating misunderstanding and wasted effort, the review process creates a more helpful and effective working atmosphere. It acknowledges the professional status of the engineer and recognizes his importance as a member of the engineering department.

4. Follow your instructor's directions and write summaries and abstracts of any additional articles you are assigned.

Informal Reports

Everybody in business—well, nearly everybody—has to write reports. They may be only a few checkmarks and comments on a form, they may be memos, or they may be book-length tomes in fancy bindings. But they will be reports, and they all fill a need for information. Whether long or short, formal or informal, they are vital means of communication among the members of a business organization.

If they're good reports, they'll help solve problems and save money for the company, and they may win points (and perhaps promotions) for their writers. If they're bad reports, they will lead to faulty management decisions growing out of inaccurate or misunderstood information. (And they may well get the writer fired.) Managers make decisions based on the information they receive in reports, and they have a right to expect that a report will be up-to-date, accurate, clear, concise, and objective. In a study published in *The Journal of Business Communication,* Hildebrandt and associates reported their findings that business communication, including report writing, was the single most useful area of study for those wishing to work in managerial positions.[1]

In this chapter, we'll take a look at the shorter, informal reports that are the lifeblood of most businesses. There are dozens of different kinds of reports, and there isn't any way we can give you a recipe for every reporting situation you might possibly encounter. Our purpose instead will be to give you tools you can use in whatever reports you may write in the future. We'll concentrate first on the qualities most reports have in common, before we look at some specific kinds of reports you are most likely to encounter.

Most of these short reports are for internal use and are, therefore, written in memo form. Those reports intended for clients or customers or other outside agencies may take the form of letters. In either case, such short reports tend to target a single reader or perhaps a small group of interested readers. They may range in length from a few paragraphs to about five pages. But they are not considered formal reports, even if they go beyond that length, unless they include the apparatus of a formal report (the transmittal letter, table of contents, appendixes, and so on).

PAFEO and the Informal Report

Everything we've discussed in Part I of this book becomes practical when you write a report. Let's see how the PAFEO process applies to the job of writing a short report.

Purpose

You can't write a good, useful report unless you know it will be used by the recipient. "Set up a shared goal," advises Linda Flower. "Sometimes a shared goal is something as intangible as intellectual curiosity. But it is the writer's job, in whatever field, to recognize goals or needs that his reader might have and to try to fulfill them."[2] Since reports should be action-oriented, you need to know what is supposed to occur as a result of your report.

- Is the report meant merely *to provide needed information?*
- Does the reader expect you *to analyze the information and draw conclusions* about what it means?
- Does the reader also expect you to *make recommendations* as a result of this analysis?

---

[1] H. W. Hildebrandt and others, "An Executive Appraisal of Courses which Best Prepare One for General Management," *The Journal of Business Communication,* 19 (Winter 1982), 5–15.
[2] Linda Flower, *Problem-Solving Strategies for Writing,* 2nd ed. (San Diego: Harcourt Brace Jovanovich, 1985), p. 162.

In the first instance, you will be writing an informative report consisting exclusively of your findings based on observation, interviews, and research.

But if your reader needs your interpretations and suggestions for action based on your findings, you will write an evaluative (or persuasive) report that includes not only your findings, but also your conclusions and recommendations.

Findings, conclusions, and recommendations: these are the three kinds of information reports generally provide. You will be able to focus on your purpose better if you have these categories clear in your own mind and label them so they are clear for the reader.

Keep the distinctions clear: *findings* are descriptive information; *conclusions* involve evaluations of what happened—reasons, value judgments, forecasts; and *recommendations* evaluate what should be done immediately or in the future.

In report writing, the P in PAFEO should also remind you to be clear about the organizational problem your report is supposed to deal with, which is separate from the rhetorical purpose you have in writing the report. If that sounds self-evident, be patient with us: experience has shown that focusing on the problem is not as easy as it sounds. In our eagerness to get to the heart of the matter, most of us display an all-too-human tendency to jump to conclusions before the problem is sufficiently clear. Kepner and Tregoe, management consultants, suggest these basic questions as ways of getting a handle on the problem. (Note how similar the questions are to the strategy of probing questions, which we discussed in Chapter 3):

1. *What* is the problem?
2. *Where* is the problem?
3. *When* is the problem?
4. *How extensive* is the problem?[3]

This is a good time to remind you of the importance of writing a good purpose statement and putting it near the beginning of the report. Remember the advice in Chapter 2, which we'll summarize here:

1. Be sure you understand the context—the organizational problem that created the need for the report.
2. Be sure you understand the specific tasks that arise out of the problem and need to be investigated.
3. Be sure you communicate to readers what you intend the report to do in relation to the problem and the specific tasks it implies.

Audience

Everything we've said about audience in Chapters 1 and 2 could be restated here. It might be worthwhile to go back and read those chapters again.

[3] Charles H. Kepner and Benjamin B. Tregoe, *The Rational Manager* (New York: McGraw-Hill, 1975), p. 45.

While we are convinced that the PAFEO process can help you with virtually any writing task, it proves especially useful in report writing.

Most short, informal reports are aimed at a single reader, or perhaps a small group of readers, and this limited audience has specific needs to be met. Since you usually know your reader(s), report writing offers a splendid opportunity to adapt to your audience. Here are some questions that may help:

- *What does your reader need to know?* Do you need to provide background before the reader can understand your main message? In many cases, you will find that the decision maker needs to know the problem, the possible solutions, and the one you recommend (which is really just another way of saying *findings, conclusions,* and *recommendations).*
- *What attitude will the reader have toward your message?* As we have discussed in Chapter 2, the way the reader feels about your message may affect your choice of inductive or deductive organization.
- *What are the reader's problems and goals?* It takes some imagination to get inside your reader's consciousness and try to see your report as it looks from that particular reader's viewpoint. But if you want your report to be psychologically effective as well as logically effective, this approach will pay off. This approach is an application in writing of psychologist Carl Rogers' suggestions for improving communication between two speakers.

  Briefly, Rogers' technique goes like this: Before you present your position, you must be able to state your listener's position back to her or him in such a way that he or she agrees to it. Since you can't do that in writing, you have to play both roles yourself and try to walk in your reader's moccasins as well as in your own. If you genuinely want to communicate, start by trying to demonstrate that you understand the reader's problems and needs well enough to see that there are goals you both share.

Format

Use the format of the report to help the reader by highlighting the significant elements. Since readers can't hold large amounts of information in mind, they expect to find internal and external clues in the writing to help them recognize the key ideas and important relationships. Such clues may be verbal or visual.

The *verbal* clues include using topic sentences to begin paragraphs, using transitional words to point out relationships, repeating important words or ideas, and using summarizing words in appropriate spots *(finally, to sum up, in conclusion).*

The *visual clues* include:

- Punctuation
- Typography
    different typeface
    italics
    numbering

    underlining

    bullets

- Arrangement on the page

    indentation

    additional white space

(and especially)

- Headings and Subheadings

The format can serve to preview your meaning for the reader (as in introductions, headings, purpose statement), guide the reader by means of internal and external signposts (transitional words, punctuation, and layout), and summarize your message (in ends of paragraphs, a conclusion, and an executive overview).

Evidence

Using the techniques we talked about in Chapter 3, you can find the kind of evidence you need to get the result you want from this particular audience. To make your case as strongly as possible, you want evidence that is adequate, representative, and relevant to the problem at hand.

Consider these questions also. Do you need statistics, tables, graphs, or other visuals to back up your presentation? Which materials can best be presented by visuals rather than by a narrative? Which sort of visual would be most effective? Chapter 9 will help you sort out the answers to such questions.

Organization

Most reports travel up the corporate ladder. The higher they go, the less likely they will be read in full. Therefore, the abstract or executive summary takes on great importance. We think it's so important that we devoted Chapter 10 to abstracts and summaries.

We suggest that you begin your report with a summary, unless the report is shorter than three pages. That initial summary should identify the problem being discussed and give your major findings, conclusions, and recommendations. The summary is a miniature of the whole report. While it is written last, it is placed first. Accompanied by a summary, your report will be more immediately useful to management because your main points are immediately communicated to the reader. In addition, the initial summary also prepares readers' expectations so that they are ready to receive what is to come and are better prepared to understand it.

Don't be hesitant about using headings in your summary, as well as in the report itself. For example, the title would be

Summary (or Executive Overview)

The main headings would be

Purpose and Scope
Findings
Conclusions
Recommendations

In reports too brief to include a separate summary, the introduction may consist of a concise summary of the conclusions and recommendations of the report, as well as necessary background on the subject and details on the methods and procedures used in obtaining your findings.

The more specialized the report, the more background you may need to provide for the reader who is a generalist, not a specialist. In memo reports to an immediate supervisor, the background may be unnecessary. Generally, the more readers you expect to have, the more you need the background information.

The usual parts of the introduction, then, include, the subject and purpose of the report; background information, if necessary; and a summary of conclusions and recommendations. In a short report, the introduction might accomplish these purposes in three or four sentences.

The body of the report is the longest part. It is a more or less detailed account of your findings on the subject. The amount of detail needed depends on the complexity of the problem, the purpose to which the report will be put, and the needs of its audience. Don't forget to use headings and other formatting techniques to make the material readily accessible to readers with differing needs.

The conclusion of the report summarizes the outcome of your investigations and analyzes the significance of what you have found out.

The recommendations, if the report contains them, are clear statements of what steps should be taken now or in the future to avoid or remedy the problem.

Kinds of Reports

What follows consists of some specific advice and samples of several kinds of reports commonly used in business. Of the many types, we have chosen minutes, progress and trip reports (typifying primarily informational types), and feasibility and incident reports (typifying evaluative emphasis). Our advice is neither absolute nor exhaustive. Every business and every boss may vary in what they expect in a particular report. As a matter of fact, businesses do not always use the same names for reports; a progress report in one company may be called a status report in another.

The best advice we can give is general: when you are asked to do a report, find out what the boss and the company expect from the report. It may be exactly what we describe here, but don't make any assumptions.

Use the PAFEO process to help you think your way through what must be done. The great advantage of the process approach to writing is that it prepares you for the unknown. You cannot ever have a model for every writing assignment you will encounter, but you don't need to follow a model slavishly if you have cultivated a problem-solving approach to writing, using the techniques we have discussed in Part I of this book.

We recognize from our own experience that many students and teachers feel more comfortable with models of reports to supplement the explanations of process. The advice and examples presented in this section are not meant to be definitive; they are simply illustrative. We have deliberately tried to keep our advice general, except when experience has shown that a particular fault often occurs in a certain kind of report. The comments are aimed at meeting the needs we have seen in our classrooms and in our business workshops.

Progress Reports

Sometimes referred to as status reports or periodic reports, progress reports keep managers informed by telling them whether the project is on time, whether it is within its budget, or whether it is encountering problems. Managers can allocate resources using the information (and possibly recommendations) received in such reports.

The typical progress or status report includes:

- Identification of the project and the time period covered by report.
- Significant problems and actions taken.
- Work completed.
- Work scheduled for the next time period.
- Conclusions and recommendations (if needed).

Generally, we recommend putting the most important information first: managers need to know about problems encountered and actions taken. Here's a progress report that follows an organizational structure outlined in a company handbook. Does it cover the information listed above?

TO: John Dempsey

FROM: Bob Welsh

DATE: November 22, 1986

SUBJECT: Weekly Status Report on Conway Co. Project

SIGNIFICANT PROBLEMS/ACTION TAKEN

While we did manage to stay on schedule this week, a
problem came up that could produce serious delays if it is

not solved. It's a touchy problem because it has to do with a key person at Conway. The head systems analyst delays responding to even the simplest request from us. We have to spend excessive time and effort getting data we need. He has acceptance authority over the entire project and might present objections at the project's completion. I have tried to talk to him about our relationship with the company, but he has made excuses and cut short our conversation. He may think that our consulting report will threaten his status or security in some way.

ACCOMPLISHMENTS

Started: testing VYSFUFOA.
Continued: coding database copy program. System was down for two hours on Tuesday, but we were able to make up for lost time, thanks to additional help from Steve and Nancy.
Completed: rewrite of report VYSFNPPG.

PLANNED ACCOMPLISHMENTS FOR NEXT WEEK

Start: testing minor changes to programs VYSFIPP1 through VYSFIPP7.
Continue: rewriting of report on engineering programs.
Complete: testing database copy program.

RECOMMENDATIONS

I think you ought to arrange to meet with the head systems analyst to see if you can determine the reason for his uncooperative attitude. If I am right about his fears for his job or his status, reassurance from someone in your position might be all that is needed to get him working with us instead of against us.

A meeting now might also give him a chance to voice any objections or reservations about the project ahead of time instead of his denying acceptance of the project at the time of completion.

Trip Reports

When you report on a business trip, you fulfill several purposes. First, you provide a permanent record of why you made the trip, what was discussed, and what was accomplished. Second, you allow other employees who did not make the trip to benefit from the information gained.

For the sake of proper filing and ease of locating, be sure that the subject line of your memo is specific about destination and dates of trip.

SUBJECT: Trip to interview Dr. E. M. Paulette, M.D., in London, Ontario, May 18-19, 1986

Use headings to divide the report into sections detailing each major event, and end the report with conclusions and recommendations if your reader expects them.

Feasibility Reports

A feasibility report tells whether a particular plan or action would be possible or desirable. A good feasibility report begins with a concise purpose statement that helps both reader and writer by clarifying the objective and scope of the study. The description of scope should also include the criteria used to measure and evaluate each alternative.

Suppose, for example, you were preparing a feasibility report to help your company decide on an office site. You would determine the main criteria to be applied and use these as major divisions, with the sites under consideration as minor divisions under each criterion. The criteria might be the market that could be accessible from each location, the cost of real estate or office rental, and the availability of transportation. These criteria would constitute your main headings, and you would then discuss the advantages and disadvantages of each alternative location with regard to each criterion. Incidentally, it is a good idea to list the criteria (the major headings) in most important to least important order. You might use a table (such as Table 11–1) to illustrate your comparison.

TABLE 11–1

	Santa Barbara	Oshkosh	Lancaster
Market (20 mi. rad.)	1.2 million	950,000	1 million
Cost	$2.5 million	$1.7 million	$1.5 million On 2 bus routes
Transportation	No pub. trans. 10 acres parking	On two bus routes 20 acres parking	Train stop 50 acres parking

The conclusion of your report summarizes the relative strengths and weaknesses of each locale, and recommends one of the locales as being the more feasible choice.

Incident Reports

Accurate, objective information is vital to a report of an incident, whether the report describes an accident, an equipment failure, or any other problem management should be made aware of. The manager needs to know the nature and cause of the problem in order to prevent it from getting worse or recurring in the future. The incident report may end up as the legal record of what actually occurred. It is most important, therefore, to keep facts (findings) distinct from inferences (conclusions).

Use probing questions to develop your information:

Who? Who was involved? Who was told first? Who was there? Who else might know something about the incident?

What? What exactly happened?

When? When did it happen? What was happening immediately before and immediately after?

Where? Where did it occur? Where were the people involved immediately before the trouble and right after it?

How? How did the events lead to the trouble? How could it have been avoided?

Why? Why did the trouble happen? What appeared to be the causes? (Be careful not to blame or condemn by your choice of tone. Let the facts speak for themselves.)

The use of the three headings—Findings, Conclusions, and Recommendations—should be helpful in organizing the incident report. The introduction of the report summarizes all three in a few sentences. The body of the report gives the findings, being precise about details that may be important if there is a legal case in the future: names, dates, times, locations, names and addresses of witnesses—any information that could be needed later. Here's a tip: imagine that your successor is reading this report four years from now; have you included enough information to make the incident comprehensible to that person? You may also draw conclusions in the body of the report about the causes of the incident. Separate headings for the findings and conclusions sections will help your readers. The final section of the report presents your recommendations about what should be done to avoid such incidents.

Minutes

Minutes are another kind of informal report. They are important because they are a permanent record of the oral discussions carried on at meetings of organizations and committees.

To be useful, minutes must include the following:

1. The name of the committee or organization holding the meeting.
2. The date, time, and place of the meeting.
3. An indication as to whether the meeting is a regularly scheduled one or one called to handle or discuss a special issue.
4. A list of who attended, who missed, and who presided at the meeting.
5. For regular meetings, a statement that minutes from the previous meeting were approved or revised, or a statement that the reading of the previous minutes was dispensed with.

6. A list of the topics discussed or reports introduced at the meeting.
7. A list of motions made that were carried, defeated, or tabled, along with an indication as to who made and seconded each motion.
8. The text of all resolutions adopted during the meeting.
9. A record of all votes (for, against, and abstain) cast at the meeting.
10. The time the meeting adjourned, and the time, place, and date of the next meeting.
11. The secretary's signature.

If you have to take minutes of a meeting, prepare for your duties by consulting a copy of *Robert's Rules of Order*. Be ready to take notes; you'll likely need more than one pen or pencil and a pad of paper. You can also bring a recorder along to tape the meeting and thereby double-check your notes when you transcribe the minutes. Write up the minutes as soon as possible after the meeting while facts are still fresh in your memory. If you have doubts about some point, consult others present at the meeting for verification.

Minutes should be brief, specific, and objective. They must be a permanent record that will make sense both to those present at the meeting and to those who were absent from the meeting but who nonetheless need to know what happened. Make generous use of white space and headings in your minutes. Make sure all references to people and items of business are clear and consistent. When referring to people, do so by full names and avoid sexism. If a man can be "Bill Smith," a woman can be "Betty Jones." Don't refer to only members of one sex by courtesy titles.

Here are minutes of a monthly meeting of unit managers of a fast-food chain:

```
MINUTES OF THE UNIT MANAGERS' FEBRUARY MEETING
King of Prussia, PA
February 8, 1986
1-3 PM


Present:   Betty Thompson      Jerry Rinoni
           Dave Bednarcik      Al Scott
           Judy Haggans        Angela Liberio
           Tina Shirer         Gene Whitry, General Manager,
                                 Presiding

Absent:    Dave Shoemaker
```

The minutes of the January 11, 1986 meeting were accepted as distributed.

1. Update on Continuing Discussions of Recurring Issues

    Gene Whitry began the meeting with an update on the following recurring issues:

a. cash discrepancy reports
b. missing deposit tickets
c. late validation reports
d. interstore transfers

2. Report on Operating Systems

Betty Thompson reported on the new procedures adopted
for operating systems:

a. Only payroll dollars for employees are to be
reported on the daily operating reports.
b. The yellow copy of the daily operating report is to
be sent to the Comptroller.
c. The Comptroller and the General Manager will
monitor all reports periodically for discrepancies.

3. New Procedure for Handling Health Inspections

Angela Liberio announced that effective immediately
all managers are to supply the home office with the
name and telephone number of all city health
inspectors.
    In addition, managers must accompany inspectors when
they make their rounds and telephone the results of
the inspection to the home office immediately after
the inspection has been completed.

4. Open Discussion

The meeting concluded with a 45-minute open discussion
about the responsibilities of the assistant managers.
As a result of that discussion, Tina Shirer and Judy
Haggans were appointed to draw up a preliminary
expansion of the assistant manager's job description
for presentation to the group at its March meeting.

The March meeting will be held in Springfield, MA, on
March 9, 1986, from 1-3 PM.

Al Scott

Al Scott, Secretary

Qualities of a Good Report: A Summary

Whatever type of report you may encounter, don't panic if it isn't one of
the reports discussed here. The general principles discussed at the beginning
of this chapter still apply. There is no rigid formula, and there shouldn't

be. But good reports have these qualities in common, as summarized by Bowman and Branchaw:

1. They are accurate, on-time, and up-to-date.
2. They are clear, concise, error-free, and helpful.
3. They are well-organized in a format that allows selective reading.
4. They are attractively presented and designed for easy readability.
5. They are cost-effective because they help solve or prevent problems for the organization.[4]

Exercises

1. Most of us use words rather loosely, and this habit can be detrimental to clear thinking and writing. In each of the following statements, indicate whether you are dealing with a finding, a conclusion, or a recommendation. Be careful not to be misled by the writer's choice of language.

   a. I have found that the combination of new employees and others nearing retirement age has caused tension and loss of productivity.
   b. I concluded that the specified input file was not the correct data file.
   c. The following are the conclusions of this committee: (1) The department should divide these invoices, setting up a separate program to handle each; and (2) there is a need to add data logic to invoice print programs to eliminate the need to update the programs each time they are run.
   d. Over 25 percent of the work force (12 employees) have been late for work three times or more in the past month.

2. Your college runs newspaper and radio advertising, with heaviest concentration in the late summer and in the period between Thanksgiving and Christmas. The advertisements are placed on the local "good music" station and in metropolitan newspapers in the nearest large cities. The advertisements draw about 8,000 inquiries, 2,000 of which translate into applications. About 800 students actually enroll as freshmen. Does the college have a problem? Using the Kepner-Tregoe questions (page 247), write a statement of the nature of the problem. Assume that the president of the college has asked you to write a report on this situation. Write an introduction to this three-page memo report, making sure that you include the subject and purpose of the report, background information if the president would need it, and a brief summary of conclusions and recommendations.

3. A frequent problem in industry results from communication difficulties between specialists and generalists. The people in the lab can't get marketing executives to understand their problems; the programmers can't communicate with the sales reps. Let's assume you are a manager who is fed up with the reports you have been getting from the computer scientists because they don't tell you what you need to know in language you can understand. Write a memo to the technical staff reporting on the problem and suggesting possible solutions.

[4] Joel P. Bowman and Bernadine P. Branchaw, *Business Report Writing* (New York: Dryden, 1984), p. 9.

4. Write a two-page memo to your boss (not your instructor) on one of the following topics:

   a. a project you are currently working on
   b. an evaluation of methods or procedures currently used by your company in a particular area
   c. an evaluation of your job performance to date
   d. a safety hazard or production snare that needs to be eliminated

5. Rewrite this report, dividing it into introduction, body (including division into findings and conclusions), and recommendations. Use verbal and visual cues to make your format more effective.

May 15, 1986

TO: Harry Sorreles, General Manager

FROM: James J. Muldoon

SUBJECT: Evaluation of Manufacturing and Quality Control
           Procedures at Orange County Plant

On May 1, 2, and 3, 1986, Jim Campbell (Production Supervisor), Al Conway (Technical Director), and I visited PennTemp's Orange County plant. The purpose of this visit was to investigate reports of incidents of poor quality of the products manufactured and shipped from this facility. Focus was on personnel, manufacturing set-up and procedures, and quality control.

The following is a synopsis of observations and recommendations:

It was observed that people and equipment are adequate from an experience and capability standpoint to produce consistent quality products. There were, however, a number of significant areas which, if addressed, would reduce the likelihood of manufacturing problems.

General Housekeeping, particularly in the blending area, was less than desirable. The whole area could use a good cleaning.

Piping system in blending area is unduly complicated. Some pipes carry two or more oils. Oil and additive delivery lines from tank storage to mixing kettles are not clearly marked. The system as it exists invites errors. At the very least, the lines and valves should be better marked and labeled.

Kettle-cleaning techniques are poor. Oil blends are made in the same kettle as water-based products. This is a sure invitation to water contamination of petroleum products. If this practice must continue, proper instruction on

```
cleaning techniques should be given and strictly adhered
to. It is hoped that the installation of a new 5,000-
gallon kettle will eliminate the need to mix oil and water
products in the same kettle.
```

6. Write a summary to be placed at the beginning of the report in Exercise 5. Assume that Mr. Sorreles may not read the entire report but will depend on the summary to give him a quick picture of what is wrong.

7. If you are a member of a committee or a student organization, write the minutes of one of your meetings. Check to see that you have included all of the relevant data as listed for you in this chapter.

8. You hold a part-time job as teller in a bank located a few miles from your campus. The manager learns that you are taking business writing in college and thinks it might be a good idea if all customer service representatives were to have this course. She asks you to write a report that will clarify whether or not this would be feasible. If feasible, would you recommend this course as appropriate for those involved? Here is some information you may need to do the report. Consider using a table to present the facts in the most appropriate way.

Course title: Writing for Business and Industry
Contents: Teaches the writing process in a business environment. Topics include informative and persuasive messages, reports, proposals, visuals.
Costs: Texts averaging $20. Tuition: $450 (day); $350 (evening).
Times: Sections offered at 8:00 and 9:00 in the day on Mondays, Wednesdays, Fridays. Evening sessions at 6:00 on Mondays and Wednesdays and also on Tuesdays and Thursdays.
Location and access: Ten minutes from work to campus by car. Parking available on campus. Bus route 100 connects bank location with campus.

CHAPTER 12

Formal Reports

The word *report* is simply a generic term for any document that is informational, analytical, or persuasive. A report may be in response to a specific request, or it may originate from your own desire to inform, analyze, or persuade. Business and industry regularly require reports, and generally, these reports demand more preparation on your part than letters or memos.

In the previous chapter, we discussed a variety of informal reports. In this chapter, we will expand our earlier discussion to include some guidelines for the preparation of formal reports.

One of the things that may set formal reports apart from the reports we have already discussed is the amount of work such reports require. Formal reports may require weeks, if not months, of work. They may be the work of one writer or of a committee of writers, each of whose members held individual responsibility for some part of the project that produced the report.

For mal reports aim at major organizational changes. They may involve expenditures of large amounts of capital. They may be part of detailed projects carried out for clients in business, industry, or government.

Formal reports may be quite elaborate in terms of format and organization. Since formal reports are usually the result of a great deal of hard work

261

and effort, you should want to present your findings as clearly as possible. Everything we have said earlier about the relationship between readers and writers applies when you write a report. Since formal reports often have multiple readers, you must consider the differing needs of your readers in a document that may run from 1 to 1,000 pages, depending on the report.

While formal reports are similar in some ways to the long papers you may have had to write in school, they are not the same as library research papers. Formal reports require more than a trip or two to the library to consult the card catalog or the *Reader's Guide to Periodical Literature*.

Formal reports can involve research or library work, but they are more likely to involve the collection and analysis of data, interviews with people, samplings of public opinions, and just plain digging and hard work.

The PAFEO Process and the Formal Report: An Overview

Purpose

To write a useful report, you must first determine how your audience will use that report. You purpose in writing will be contingent upon your audience's purpose in reading. In addition, then, your purpose is tied to the problem you are attempting to address in your report. In discussing informal reports in the previous chapter, we used an approach suggested by Kepner and Tregoe that is helpful in defining purpose. This approach, which works equally well for defining purpose in a formal report, requires you to ask yourself—and your audience—a series of questions designed to describe the dimensions of the problem that your report seeks to address.

1. *What* is the problem?
2. *Where* is the problem?
3. *When* is the problem?
4. *How extensive* is the problem?[1]

Defining the key problem the report addresses is crucial to the succsss of your report. To clarify your purpose in writing a report, you need to define the problem as precisely as you can.

Audience

Readers see more reports than they read. A formal report presents you with a slightly more complicated audience problem than a letter or a memo. Pace, tone, and readability are still important, but the audience for a report may be broader than that for a letter or memo.

[1] Charles H. Kepner and Benjamin B. Tregoe, *The Rational Manager* (New York: McGraw-Hill, 1975), p. 45

The more formal the report, the more people there are who are likely to see it. A corporate annual report is a kind of formal report. Its audience is extremely broad-based and includes shareholders, would-be shareholders, corporate employees, stock brokers, market analysts, watchdog groups and agencies, government regulatory agencies, and a host of others.

As they travel up and down the corporate ladder, your formal reports will cross many desks. Few of your readers will plow through every page. The higher up the corporate ladder your report travels, the less likely it is to be read from cover to cover. Top management will only want—and have time to read—an effective summary. Those further down the corporate ladder—people who are also busy—may need to pick and choose among your contents, finding only those sections of your report that are useful to them.

Ironically, the entire report may not be read until long after you have written it. Your report may lie buried in a file cabinet or gathering dust on a shelf until someone doing research for another report comes along looking for evidence. Our own university conducted a self-study in 1985 in preparation for a regular ten-year reaccreditation visit. One of the more important documents in that 1985 study was the final document that emerged from the 1975 study. That 1975 report had probably gone unread for most of the previous decade, but in 1985, it had a new life as an important document when addressing a current problem.

Reports are often directed at specific audiences, but they are just as often reproduced and circulated to interested readers. Format and organization will then be especially important to you when you write a report. Format and organization can help you address the different needs of different readers who may be reading your report at different times or under different circumstances.

Format

Some companies have set formats for formal reports. Other companies allow writers wide latitude in formatting reports. Whether your format is set or self-determined, remember that it should help you achieve your purpose with your audience.

While all reports have a beginning, a middle, and an ending, formal reports may be further broken down into more specific parts. These parts can include the following.

A LETTER OR MEMO OF TRANSMITTAL   A letter of transmittal is addressed to an external audience, while a memo of transmittal is addressed to an internal audience. Both explain how, why, and under what circumstances your report was prepared.

A COVER   Formal reports require professional, formal presentations. A cover includes such information as the name of the organization and division, the title and date of the report, and the names and titles of the author or authors. Depending on the circumstances surrounding your report and its presentation, your cover may be clear plastic or gold-embossed leather.

A TITLE PAGE   A separate title page includes the following data: the title of the report, the names and titles of the recipient and of the authors or authors, and the date of the report. The title page can also include some information about the circumstances that lead to the writing of the report or the context in which the report will be read. Academic dissertations— a kind of formal report—often include a statement on their title page similar to the following:

> A Dissertation Submitted in Partial Fulfillment of the Requirements for the Doctor of Philosophy Degree in Economics at Standford University.

Likewise, a periodic study of a problem might include on its title page the information that it was

> A Quarterly Reports on Earnings and Losses in Amalgamated Federal's Eastern and Southern Subsidiaries.

A SUMMARY   Every formal report requires a summary covering purpose, scope, findings, conclusions, and recommendations. The summary may be the most widely read part of your report, in many cases, it may be the only part of the report that is read with any care.

TABLE OF CONTENTS   The more complicated your report, the harder the time your reader will have in finding whatever he or she wants in it. The longer your report, the more formidable it will appear to your reader who may be disinclined even to look at, let alone read, the entire document.

A table of contents carefully tied to the parts and the headings of your report will provide your reader with ready access to needed information. If your report contains visuals and an appendix, don't forget to include them in a separate table of contents.

INTRODUCTION   Some reports present information that is immediately accessible to readers; others must first prepare readers for what is to follow. A letter or memo of transmittal, a summary, and a table of contents all help to prepare a reader for what is to follow. Sometimes, however, your reader will need more in the way of preparation than these parts of a formal report provide.

An introduction can help you meet your reader's need for more preparation by providing background data, details about your approach or methodology, limitations to your study, criteria you used in conducting research or in making evaluations.

In technical or engineering reports, you can also use your introduction to list, define, or translate any abbreviations, terms or symbols that the reader must know before he or she can fully understand your report.

BODY   The body of your report will consist of the text and visuals you present to your readers. Page after page of solid print is always deadly. Since reports can run into the hundreds of pages, your reader will rely upon the headings you provide in order to distinguish among the points and topics you present. In some cases, every paragraph may need a heading; in others, you may simply need a heading for each of the major sections of your report. Remember, though, that both a typewriter and word processor present you with a variety of ways of incorporating major and minor headings and subheadings into the body of your report. Such headings help your readers and yourself keep track of the logical divisions in your report and of the relationships the parts of your report have to one another.

CONCLUSIONS   All reports should have conclusions in which you tie together the information presented in the body of the report.

RECOMMENDATIONS   Reports that do more than simple fact finding—reports that suggest a course of action—should contain a set of recommendations.

APPENDIX   An appendix is useful for presenting statistics, tables, and other information of secondary rather than primary importance to your readers. When your report is built around visuals, incorporate your visuals into your report's body. When visuals are simply supportive of or tangential to your report's body, move the visuals into an appendix.

If you list your references as endnotes, rather than as footnotes placed on appropriate pages, your endnotes can comprise an appendix to your report. A bibliography of works consulted or of works for further reading can comprise another appendix. Sample questionnaires or model forms, formats, or other materials can comprise still another appendix.

Be careful, though; an appendix is not designed to be a trash heap for whatever is left over from the body, or other parts of your report.

Evidence

The amounts and kinds of evidence—verbal and visual—that you will need will vary from report to report. As you compile your evidence, remember to distinguish among findings (descriptive information), conclusions (evaluations or judgment about what is happening), and recommendations (steps that need to be taken).

Organization

Your organization, like your prose, needs to be reader-based, not writer-based. An organizational pattern that you use in writing your report (say, simple chronology) may not present information to readers in the most helpful of ways. A busy reader may need that information organized in terms of cause and effect.

Reports like any other document will have a beginning, a middle, and an end, but there is no reason that these sections need to appear in that order; your conclusions or recommendations might, for instance, come first. Armed with a table of contents, an informative cover letter or memo, and a clear set of headings and subheadings, you can present your evidence to your audience in a variety of organizational patterns. We provided you with details on these patterns earlier in our general discussion of organization in Chapter 4.

Audience, Format, and the Formal Report: Some Additional Comments

The longer or more complicated your report, the greater the number of problems you can create for your readers. Like all audiences for business and technical documents, readers of your formal reports are busy. They expect your formal reports to have a purpose and an arrangement which is clear, and they know less about the subject of the formal report than you do.

In the previous section of this chapter, we discussed the various parts a formal report can have. Three of these parts—the summary, the introduction, and the letter or memo of transmittal—can be especially helpful in solving the problems formal reports present to your readers. Elements of format can, once again, solve problems of audience by providing your readers with as many guideposts as possible as they skim or read through your report.

The Summary

In Chapter 10, we discussed summaries and abstracts in great detail. Here, we need simply to reemphasize the fact that *all* formal reports should contain a summary. We know of one local company where all documents one typed page or longer must begin with a summary. Such an approach risks overusing the summary on one hand, but, on the other hand, it recognizes that readers inside and outside that company confront more documents on a daily basis than they are able to read fully and carefully.

The Introduction

An introduction can help your readers understand and digest information contained in the body of your formal report. A good introduction does more than simply "open the report." Instead, it can:

- State the subject of the report.
- Indicate the purpose of the report.
- Warn the reader about any limitations to the report.
- Provide any background information helpful to the reader.

- Define any terms crucial to the report but unfamiliar to the reader.
- Preview the report's method of development so that the reader has an indication from the start of not only *what* the report discusses but also *how* the report discusses.

Introductions can be fairly detailed, but they are not summaries. A summary is a self-contained synopsis of the entire report. An introduction is simply designed to lead the reader into the body of the report. The less expertise your readers have about the subject of your report, the more you will have to depend on your introduction to help narrow the gap between what they and you know. Because readers can be so dependent on introductions, you need to organize them as carefully and as clearly as possible.

Here are two sample introductions. As you read them, keep asking yourself if the introductions help or hinder your progress in reading or digesting information that is to follow.

The first introduction is from a corporate annual report. The audience for such a report is, as we indicated earlier, large and varied in its levels of expertise and interest, but each reader wants to know a common piece of information, the financial condition of the company. Here's the introduction:

> We are pleased to give you this report on 1986. Our company demonstrated considerable strength during a troubled economic year around the world, and we took several key steps toward exploiting the exciting new opportunities that technology is creating for our company.
>
> Revenues rose past the $4 billion mark for the first time, although earnings were slightly below last year's. We were successful in improving what was an already strong balance sheet.
>
> In a difficult economic environment, the major operating units of our company competed well. We continued to benefit from the balance we achieved through long-standing commitment to diversification. In 1986, we moved further in that direction as we began applying new technologies to the communication of entertainment and information.

Confused? Ready to sell your stock, or to buy more?

The introduction fails for a number of reasons, the most important of which is that it fails to introduce anything. Instead, it hits the ground running and leaves the reader wondering where the report is going and what the introduction is trying to say.

This introduction is in some ways a cross between an introduction and a summary, although in the final analysis it is neither. These paragraphs fail rhetorically. They do not do what readers expect an introduction to do. Even worse, the paragraphs may lead the reader to question the financial stability of the company. In what they fail to say, these paragraphs may lead the reader to believe the company is hiding something. At best, this introduction will be of no help to the reader; at worst, it may annoy, anger, or worry the reader since the company seems to be talking around the topic rather than about it.

The reader could care less that management is pleased to send along the report. Instead, the reader wants to know why management is pleased and then some general points about the company's performance that can be expanded upon in the body of the report. The reader wanting a preview of what is to come finds what seems to be doubletalk instead.

Here's another introduction, to IRS Publication 463—Travel, Entertainment, and Gift Expenses. Notice the different approach the writer has taken in this introduction, which incidentally is clearly labeled in boldface print INTRODUCTION.

> This publication explains business-related expenses for travel, transportation, entertainment, and gifts that you may deduct on your income tax return. It also discusses the reporting and record-keeping requirements for these expenses.
>
> Following a summary of the general requirements for deductibility by employees, self-employed persons (including independent contractors), and employers (including corporations and partnerships), the publication presents information on the rules governing these business expenses in terms of eligibility, proof, and reporting. Use the index to help locate discussions of general topics that may apply to you.

There is no denying that tax regulations are complicated, but the explanations of these regulations need to be easy to follow. While we won't vouch for the readability of the contents of Publication 463, we do think the example does an excellent job of what an introduction should do: it tells what the publication is about and how the publication discusses what it discusses.

The Letter or Memo of Transmittal

Written last, but read first, a letter or memo of transmittal recognizes the fact that your readers don't want your formal reports arriving on their desks without some hint of why they are there.

On the simplest level, this letter or memo says "Here is a report I have conducted of Y," or "Here is the report you requested on X." But the letter or memo of transmittal can do more to help your readers and yourself in getting the report read.

A letter or memo of transmittal can do more than simply announce the fact that a report follows; it can also prepare the reader for what is to follow. Such preparation helps you, the writer, too, since it can lead your readers to do what you want them to do—read and act upon your report. A letter or memo of transmittal can

- Define the topic of the report.
- Alert the reader to any special features of the report.
- Preview the report's conclusions and recommendations.
- Stress the importance of the report.
- Encourage the reader to follow through on the report.

Here is an example of a letter of transmittal that does more than simply say "Here's the report."

Dear Professor Soven:

    Here is the formal report you requested surveying the
technical writing responsibilities of an entry-level
engineer.
    The report presents data based upon the engineering
practices of Sanders and Thomas, Inc., a consulting firm
in Bossier City.
    These data are presented in several parts including a
survey of the general requirements for employement in an
entry-level position with Sanders and Thomas, the results
of several interviews with managers and officers at
Sanders and Thomas, and an analysis of several writing
samples supplied by Mr. Paul Retz, Director of Personnel
for Sanders and Thomas.
    The appendixes to this report contain several of these
writing samples for your further reference.

Such a letter of transmittal invites the reader to continue reading the report
by providing yet another guidepost that shows the writer recognizes just
how busy the reader is.

Evidence and the Formal Report

We live in an age of incredible advances in knowledge and technology.
While the formal report is not the same thing as a library research paper,
the library remains a major source for information. If your formal report
does require library research, talk with the staff at your local, school, or
corporate library about the references that are available to you.

Basic References for the Business Student

What follows is a list of basic references, a place to start, for students
and researchers in the areas of business and industry. (Our thanks to the
reference staff of the David Leo Lawrence Library at La Salle University
for allowing us to reprint this list.)

To find magazine and newspaper articles, use:

Business Periodicals Index. This is a cumulative subject index to English language
    business periodicals. Following the main body of the index there is an author
    listing of citations to book reviews. The index is published monthly except
    August, with a bound cumulation each year.
Wall Street Journal Index. This index to the Wall Street Journal is published
    monthly. A permanent bound volume is published each year. The index is in
    two parts—the first section is *Corporate News,* the second, *General News.*
    Each entry includes a brief abstract of the article.

F & S Index of Corporations and Industries. This index covers company, product and industry information from over 750 financial publications, business-oriented newspapers, trade magazines, and special reports. It is in two parts—the first section covers *Industries and Products,* the second covers *Individual Companies.*

Public Affairs Information Service. This is a selective subject listed in the areas of economic and social conditions, public administration and international relations. This index, in addition to selectively indexing journals, also covers selected books, pamphlets, government publications, and reports of public and private agencies.

For up-to-date information on a company or a specific industry, use:

Moody's Manuals. These manuals cover U.S., Canadian, and other foreign companies listed in the U.S. Stock Exchanges. Information about each company usually includes a corporate history, officers and directors, balance sheet statistics, etc. Volumes are published annually with semi-weekly supplement updates.

Standard and Poor's Industry Surveys. Revised triennially, this two-volume reference work is a source for basic data on 69 major domestic industries. For an analysis of trends and outlooks this is a useful tool. Included are statistics on sales and earnings, profit margins, price-earnings ratios, etc., of leading companies in the industry.

Dun & Bradstreet. *Million Dollar Directory. Volume I* lists approximately 39,000 U.S. companies with an indicated worth of $1 million or more. Gives officers, products or services, SIC number, approximate sales and number of employees. *Volume II* gives similar information for U.S. companies with an indicated worth from $500,000 to $999,999.

Standard and Poor's *Register of Corporations, Directors, and Executives.* Published annually, lists officers, products, SIC number sales range, and number of employees for approximately 36,000 U.S. and Canadian companies.

Disclosure, Inc., *Annual Reports and SEC 10-K Reports* for the Fortune 500 companies.

For business and industry forecasts, consult:

U.S. Dept. of Commerce. Industry and Trade Administration, *U.S. Industrial Outlook* for 200 industries with projections. Published annually, this volume presents information on recent trends and the outlook for over 200 individual industries.

Predicasts, Inc. *Predicasts.* This is a quarterly service covering short- and long-term forecast statistics for basic economic indicators and for individual products.

For current business statistics, see:

Federal Reserve Bulletin. Published monthly, this is the best source for current U.S. banking and monetary statistics.

Business Conditions Digest. Monthly. Charts and back-up statistical tables for those leading economic time series of most use to business analysts and forecasters.

Monthly Labor Review. Best source for current statistics on topics such as employment, consumer and wholesale prices, and productivity.

CPI Detailed Report. Monthly report on consumer price movements.

Economic Indicators. Statistical tables and charts for the basic U.S. economic indicators. Published monthly.

Survey of Current Business. The most important single source for current business statistics.

Important reference sources are:

U.S. Bureau of the Census. *Statistical Abstract of the U.S.* This annual reference work serves as the prime source for U.S. industrial, social, political, and economic statistics. Usually covering 1 to 2 years, it is updated by such publications as the *Survey of Current Business.* For historical statistics consult the supplementary volume, *Historical Statistics of the United States, Colonial Times to 1970.*

U.S. Department of Commerce. *Business Statistics.* Biennial supplement to the *Survey of Current Business.* This publication presents historical data for approximately 2,500 series that appear in the "S-" pages of the *Survey of Current Business.*

The Annual *Statistical Yearbook of the United Nations* serves as a summary volume of the *International* statistics currently available.

Computer-Assisted Research

A rapidly increasing number of online databases provide quick access to a great deal of information. Online searches are possible in almost any discipline. Such searches vary in cost depending on the amount of computer time used and the number of citations produced. The hourly fees for these searches can run as high as $100.00, but the searches are conducted so quickly that an average search should cost you no more than $30.00.

A selected list of available databases follows. For more comprehensive lists, see

Doran Howitt and Marvin L. Weinberger, *Inc. Magazine's Data Basics: Your Guide to Online Business Information* (New York: Garland, 1984).

Bernard S. Schlessinger, *The Basic Business Library: Core Resources* (Phoenix: Oryx, 1983).

An experienced librarian can help you plan the terminology and strategy most likely to provide a useful bibliography at a reasonable cost.

Business

Company and Financial

Disclosure II provides extracts of reports filed with the SEC by publicly owned companies.

D & B—Dun's Market Identifiers presents detailed information on more than two million U.S. businesses with ten or more employees.

D & B—Million Dollar Directory comprises comprehensive business information on 115,000 U.S. companies from the three-volume *Million Dollar Directory Series.*

Electronic Yellow Pages—Financial Services Directory provides online yellow-page information for banks, savings and loan institutions, and credit unions in the United States.

Electronic Yellow Pages—Manufacturers Directory provides online yellow-page information for manufaturing establishments in the United States.

Electronic Yellow Pages—Professionals Directory provides online yellow-page information for professionals in insurance, real estate, medicine, law, engineering, and accounting.

Electronic Yellow Pages—Retailers Directory provides online yellow-page information for all types of retail businesses.

Electronic Yellow Pages—Services Directory provides online yellow-page information for all types of services including financial services, business services, and office and recreational services.

Electronic Yellow Pages—Wholesalers Directory contains listings of all types of wholesale dealers from 4,800 telephone directories throughout the United States.

International Dun's Market Identifiers contains directory listings, sales volume and marketing data, and references to parent companies for over 500,000 non-U.S. companies.

Investext provides full-text financial research reports from 27 of the leading investment banking firms in the United States, Europe, Canada, and Japan.

Moody's Corporate News—International offers comprehensive coverage of business news and financial information on more than 3,900 major corporations and institutions in 100 countries worldwide.

Moody's Corporate News—U.S. offers comprehensive coverage of business news and financial information on approximately 13,000 publicly held U.S. corporations.

PTS Annual Reports Abstracts provides comprehensive coverage of annual reports issued by over 3,000 publicly held U.S. corporations and selected international companies.

Standard & Poor's Corporate Descriptions offers in-depth descriptions of more than 9,000 publicly held U.S. corporations, including a complete corporate background, all income account and balance sheet figures, and important stock and bond data.

Standard & Poor's News offers both general and financial information on more that 10,000 publicly owned U.S. companies.

Standard & Poor's Register—Corporate provides important business facts on over 45,000 leading public and private U.S. and non-U.S. companies.

Management and Economics

ABI/Inform covers all phases of business management and administration.

Economic Literature Index provides an index of journal articles and book reviews from 260 economic journals and from approximately 200 monographs per year.

Foreign Trade & Econ Abstracts provides coverage of the world's literature on markets, industries, country-specific economic data, and research in the fields of science and management.

Management Contents provides current information on a variety of business- and management-related topics to aid individuals in business, consulting firms, educational institutions, government agencies or bureaus, and libraries in decision-making and forecasting.

Computer Science

Business Software Database describes over 3,500 software packages with business applications for use with micro- and minicomputers.

Computer Database contains abstracts from 530 journals, newsletters, tabloids, proceedings, and meeting transactions, as well as business books and self-study courses, covering almost every aspect of computers, telecommunications, and electronics.

Menu—The International Software Database provides a comprehensive listing of commercially available software packages for any micro- or minicomputer.

Microcomputer Index provides a subject and abstract guide to magazine articles from 50 microcomputer journals.

Law and Government

ASI provides a comprehensive index of the statistical publications from more than 400 central or regional issuing agencies of the U.S. government.

Federal Research in Progress provides access to information about ongoing federally funded research projects in the fields of physical science, engineering and life science.

Legal Resource Index provides cover-to-cover indexing of over 750 key law journals, six law newspapers, and selected legal monographs.

NCJRS covers all aspects of law enforcement and criminal justice.

Science and Technology

Compendex provides abstracted information from significant engineering and technological literature worldwide.

GEOREF provides access to a variety of publications in the fields of geology, geochemistry, geophysics, mineralogy, paleontology, petrology, and seismology.

NTIS provides information on government-sponsored research in energy conservation, pollution, and urban and regional planning.

TRIS provides information on research in the areas of transportation.

Miscellaneous

AIM/ARM indexes materials on vocational and technical education, job training, and vocational guidance.

APTIC covers all aspects of air pollution.

Facts on File provides a weekly record of current events compiled from worldwide news sources.

Grants provides information on more than 2,000 grant programs available through public and private sources.

Magazine Index offers broad coverage of general interest magazines.

PAIS International contains references to information in all fields of social science.

World Affairs Report digests worldwide news as seen from Moscow.

Footnotes, Endnotes, and Bibliographies

If your report involves library or other kinds of research, you must document your sources. Such documentation appears as footnotes, endnotes, and a bibliography. In addition, some reports add explanatory asides to the main text by using footnotes or endnotes rather than a full-blown appendix.

First, some definitions:

- A *bibliography* is an alphabetically arranged listing of works consulted or of works offering additional information. Bibliographies are usually placed at the end of a formal report.
- An *annotated bibliography* not only lists works but also briefly describes what the works contain or discuss. The entries in an annotated bibliography are often short *descriptive abstracts*. An annotated bibliography, such as the one presented in Figure 12–1, is especially useful to your readers when you are simply listing additional sources of information.
- An *endnote* indicates the source of material quoted or paraphrased, directs the reader to additional information, or briefly explains some specific point in the text. Endnotes are compiled chronologically at the end of the report or at the end of a chapter or report section in much longer reports.
- A *footnote* does all the things an endnote does, but it occurs at the bottom of the page. Footnotes may be more difficult for you to type—a problem easily solved by some word-processing software—but they do help the reader by eliminating the need to flip back and forth from text to endnote.
- A *reference* is both a generic term for a bibliography entry and an end- or footnote, as well as a more specific term for a modifed form of notation that includes brief forms of documentation within the report text itself that are tied to a list of works consulted or a bibliography.
- *Plagiarism* is the use of someone else's words, phrase, ideas, or data without proper acknowledgment. At best, plagiarism is unethical; at worst, it is illegal.

Bibliographical entries or full-blown end- or footnotes need to provide your readers with as complete a set of publication or reference data as possible. Typically, such data include:

Author's name
Title of work
Publication facts

For books:	For journal articles:
Edition number	Title of journal
Place of publication	Title of article
Publisher's name	Volume number
Publication date	Publication date
Volume and page numbers	Page number

Journal articles

Adams, Tom. "Developing a Meeting Memo." *Supervisory Management,* 25 (July, 1980), 39–42. The 13 essentials of an effective meeting memo.

Ammannito, Theresa A. "The Introduction." *STWP Review,* 9 (July, 1962), 11–14. An examination of the correlation between reader expertise and the length and complexity of introductions to technical reports and articles.

Arnold, Christian K. "The Construction of Statistical Tables." *IRE Transactions on Engineering Writing and Speech,* EWS–5 (August, 1962), 9–14. Guidelines for constructing statistical tables, criteria for choosing vertical or horizontal presentations, and rules for the use of captions.

Aziz, A. "Article Titles as Tools of Communication." *Journal of Technical Writing and Communication,* 4 (Winter 1974), 19–21. Suggestions for forming titles that contain as much information in as few words as possible.

Bergwerk, R. J. "Effective Communication of Financial Data." *Journal of Accountancy,* 129 (February, 1970), 47–54. Suggestions on using tables to compare several pieces of data and graphs to show trends.

Bernheim, Mark. "The Written Job Search—Doubts and 'Leads.'" *The Technical Writing Teacher,* 10 (Fall 1982), 3–7. Practical strategies for job seekers.

Biskind, Elliot. "Writing Right." *New York State Bar Journal,* 42 (October, 1970), 548–554. Suggestions for improving the quality of legal writing.

Booth, Vernon. "Writing a Scientific Paper." *Biochemical Society Transactions,* 3 (1975), 2–26. A discussion of prewriting techniques, style, visual aids, and manuscript submission guidelines.

Bram, V. A. "Factors Affecting the Readability of Scientific and Engineering Texts." *The Communicator of Scientific and Technical Information,* 33 (October 1977), 3–5. Suggestions for using effective sentences as a basis for successful communication with readers regardless of their expertise.

Bromage, Mary C. "Gamesmanship in Written Communication." *Management Review,* 61 (April, 1972), 11–15. Techniques for improving business writing diction.

Christenson, Larry. "How to Write an Office Manual." *American Business,* 25 (July, 1955), 26–27, 32–34. Nine principles for writing effective manuals.

Cortelyou, Ethaline. "Abstract of the Technical Report." *Journal of Chemical Education,* 30 (October, 1955), 523–533. A discussion of the differences between abstracts and summaries.

Darian, Steven. "Using Spoken Language Features to Improve Business and Technical Writing." *ABCA Bulletin,* 44 (September, 1982), 25–30. Suggestions on the ways the differences between spoken and written communication can be used to improve business and technical writing.

Davidson, Jeffrey. "How to Spot a Phony Resume." *Supervisory Management,* 28 (May, 1983), 40–43. Guidelines for personnel managers that can serve as a warning to job applicants when preparing resumes and cover letters.

Davis, Jeffrey. "Protecting Consumers from Overdisclosure and Gobbledygook." *Virginia Law Review,* 63 (October, 1977), 841–920. A thorough discussion by a lawyer of the problems excessive disclosure, extraneous clauses, and intractable language cause consumers when they must agree to credit contracts.

DeBakey, Lois. "The Persuasive Proposal." *Journal of Technical Writing and Communication,* 6 (Winter 1976), 5–25. A discussion of the skills needed by investigators who write research proposals.

DeBeaugrande, Robert. "Information and Grammar in Technical Writing." *College Composition and Communication,* 28 (December, 1977), 325–332. A discussion of how grammar helps writers organize information.

DeRoche, W. Timothy. "Addressing Mail in the Eighties." *Balance Sheet,* 62 (February, 1981), 219–222. General recommendations for addressing mail that conform to the preferences of the U.S. Postal Service.

FIGURE 12–1   Annotated Bibliography

Source: From *Strategies for Business and Technical Writing,* Second Edition, by Kevin J. Harty, copyright © 1986 by Harcourt Brace Jovanovich, Inc. Reprinted by permission of the publisher.

For references to items such as newspaper articles, reports, pamphlets, interviews, and speeches, you will want to supply as much information as possible to help your reader find your sources.

There is no one correct way for compiling bibliographies or for listing end- and footnotes. Most writing courses follow a format laid down by the Modern Language Association (MLA), but other formats are equally acceptable as long as they provide your readers with all the data they need. Some professional organizations as well as some companies have their own formats. Whichever you use, be consistent and try as much as possible to help your reader.

The most widely used formats for bibliographical entries and end- and footnotes are discussed in detail in the following works:

American National Standard for Bibliographic References. New York: American National Standards Institute, 1977.

Campbell, William G., and Stephen V. Ballou. *Form and Style: Theses, Reports, Term Papers.* 5th ed. Boston: Houghton Mifflin, 1978.

The Chicago Manual of Style. 13th rev. ed. Chicago: University of Chicago Press, 1982.

Gibaldi, Joseph, and Walter S. Achtert. *The MLA Handbook for Writers of Research Papers,* 2nd ed. New York: Modern Language Association, 1984.

Howell, John B. *Style Manuals of the English-Speaking World.* Phoenix: Oryx Press, 1983.

Publication Manual of the American Psychological Association. 3rd ed. Washington, DC: American Psychological Association, 1983.

Trimmer, Joseph A. *A Guide to the New MLA Documentation Style* Boston: Houghton Mifflin, 1984.

Turabian, Kate L. *A Manual for Writers of Term Papers, Theses, and Dissertations.* 4th ed. Chicago: University of Chicago Press, 1973.

————. *Student's Guide for Writing College Papers.* rev. 3rd expanded ed. Chicago: University of Chicago Press, 1977.

Increasingly, there has been a movement away from notes at the foot of the page or at the end of the chapter or report in favor of a system that incorporates either brief or full-blown references into the actual text of the report. Such brief references are, as we said earlier, then tied to a list of works consulted or a bibliography at the end of the report. More full-blown references, while interrupting the flow of the text, do provide readers with references to sources immediately after the material or data referenced.

Here are three ways of incorporating references within the text of a report using parentheses at appropriate points. Use any of these ways, or the more traditional systems of end- or footnotes, to eliminate the risk of plagiarism and to help your reader find the sources you used or the references that will offer additional information.

In recent years, several writing texts have taken a "process" approach in their discussions of business and technical writing. One such text advocates a simple, easy-to-follow approach that asks writers to consider purpose, audience, format, evidence, and organization (John Keenan, *Feel Free to Write* [New York; Wiley, 1982], pp. 21–38).

Keenan's approach (*Feel Free to Write,* pp. 21–38) asks writers to consider purpose, audience, format, evidence, and organization.

Keenan argues that his approach will help writers avoid the lack of clarity that is "the most frequently cited weakness when executives evalute their employees" (p. 21).

The second and third examples are obviously more closely tied to the accompanying bibliography than is the first.

Plagiarism

If you are a college or university student, you have doubtless been warned against plagiarism in your academic writing assignments. Plagiarism in the publishing and in the video and recording industries has been the basis for any number of lawsuits, which have resulted in stiff fines for those found guilty of plagiarism. Plagiarism can be an issue in business and technical writing as well.

Plagiarism is using someone else's ideas or words without properly acknowledging the source for those ideas or words. In its most blatant form, plagiarism is the repetition, more or less verbatim, of someone else's words or ideas without providing documentation as to source in a foot- or endnote. More subtle plagiarism can occur through paraphrase or through passing off someone else's development of an idea as if it were your own. Again, plagiarism occurs when you paraphrase or take credit for what is not your own.

To avoid plagiarism—and its penalties, which can range from raising questions about your integrity to dismissal from your job and even a lawsuit—let honesty be your guide. When you quote someone directly, set the quoted material off with quotation marks and supply appropriate documentation as to your source. When you paraphrase, simply supply appropriate documentation. When in doubt, err on the side of conservatism, and cite your source or sources. In all cases, follow a consistent style for documentation.

A Sample Formal Report

The report that follows was prepared by a management consulting service. We include it because it provides a good example of how format and organization can match evidence to audience needs and writer's purpose.

PAULCOMP, INC.
REVIEW OF PRICING, BILLING,
AND COLLECTION METHODOLOGIES

PREPARED BY
JACKSON-WILSON PARTNERS, INC.
AUGUST, 1986

SUMMARY

Purpose and Scope

As requested, we have reviewed the pricing structure
for data processing services provided by Paulcomp, Inc.,
to the parent company, Paulis Publishing, Inc., and other
subsidiaries of Paulis Publishing.

Findings

In addition to discovering serious problems in pricing
policy, our review also brought to light several additional
problems which must be addressed concurrently.

These problems are found in:

>   *billing
>   *collection
>   *cost accounting
>   *quality control.

Conclusions

Improving policies and procedures in these four areas is
a must. Not only will such improvement result in better
service to current clients, but it will also support
Paulcomp's longe-range business goal of marketing their
services externally to clients outside the corporation.

In addition to bettering service, the needed improvements
will benefit Paulcomp by:

>   1. providing consistency in service and pricing
>   2. eliminating billing errors
>   3. increasing revenue through equitable billing
>   practices.

Recommendations

We recommend that management take immediate action to
achieve increased efficiency and effectiveness in billing,
collection, cost accounting, and quality control.

These improvements should go hand-in-hand with the
development of a new pricing algorithm. The increased
confidence of users which will result from the improvements will also reduce resistance to need changes in the
pricing structure.

PURPOSE AND SCOPE

The purpose of our review of Paulcomp's pricing procedures
was to survey the consistency of their pricing methology
and to identify problems in related areas such as billing
and collection.

Our charter was primarily to gather data and make observations in an overview. Many of the conclusions listed
in the next section of this report will require additional
study to define further the methodology that should be used
to implement them.

We have organized this report in parallel columns so that
our findings and our conclusions can be more easily
compared.

We believe this study will have been a success if it serves
to make management aware of the range of problems and if
it assists them in directing their efforts to correct these
problems.

Our recommendations include a brief discussion of an
approach for a program to solve the range of problems
we have uncovered. We also propose to support Paulcomp's
management in the development of this program and in
its implementation.

In performing this study, we conducted interviews with the
following Paulcomp personnel:

    Geraldine Faber, Vice President, Systems Development

    Francis Teramana, Assistant Vice President, Systems
                Development

    Cynthia Sterling, Vice President, Operations

    Leroy Hopkins, Vice President, Technical Services.

FINDING

1. There is no standard policy for pricing to ensure that all users are being charged equitably. Some prices are based on lines printed or records processed; other are just priced per run.

2. Pricing is based on the number of updates processed in a run or the quantity of output produced such as number of printed lines. It is extremely difficult to track costs versus revenues using this system.

CONCLUSION

A standard price list should be established for all users for standard services, independent of the user being serviced or the system being run. If reduced prices are necessary for certain users, they should be based upon standard discounts or negotiated agreements.

Pricing should be based on some measure of system utilization which is generally a summation of three measures*:

1. connect time costs
2. processing (or output) costs
3. storage costs.

*We have initiated a survey of the service bureau market place for comparative analysis. We do not believe it would be beneficial to present our initial observations at this time. Rather, we recommend a more in-depth comparative analysis be performed, which takes into consideration the marketplace and the specifics of your clients.

FINDING

3. Fixed price charges for services are established before the service is performed and, in the case of system development projects, often before the service is fully defined. This practice often results in inappropriate pricing for services, generally to the detriment of Paulcomp.

4. There is no standard system, either manual or mechanized, for logging system services for billing purposes. System produced statistics and manually prepared worksheets are available in some cases, but often the Billing Supervisor receives no documented notice of services performed.

5. There are inadequate controls to ensure that

- all services are billed
- all bills are accurate
- all operating procedures are followed.

CONCLUSION

We recognize that marketing, or other business reasons, sometimes dictate the need for this practice. However, we conclude that this "general" practice should be discontinued and that "concept pricing" should be quoted only as an exception. A pricing policy and methodology needs to be developed to eliminate price quotations, except by senior management, to improve the accuracy of pricing and to limit Paulcomp's risk of loss.

An automated job accounting system should be installed to produce accurate and timely reporting of computer services rendered. Daily, weekly, and monthly reporting should be available. In addition to providing notice of jobs run, system costs and utilization statistics could also be made available to track costs.

An in depth analysis of the current controls should be conducted to develop adequate procedures and controls.

FINDING

6. The billing for Paulcomp is currently done manually on a weekly basis
even though the users pay their bills
monthly. The weekly bills are sent
to individual users with the result that
about 200 bills are being issued. These
bills are then consolidated by Accounts
Payable in the appropriate department
or subsidiary, and one check per user
is mailed to Paulcomp.

7. There is no automated Accounts
Receivable system.

CONCLUSION

Billing should be done on a monthly
basis to reduce personnel requirements
for billing. The bills themselves, should
be consolidated at Paulcomp to reduce
the amount of time currently required
for manual bill preparation. If this
change in procedure is not a feasible
method for reducing the number of bills,
than management should consider the
installation of a computerized billing
system.

An automated Accounts Receivable system
is unnecessary in the near future if
monthly charges are consolidated and if
the number of bills is greatly reduced.

However, standard Accounts Receivable
reporting is needed. Management should
develop a format for these reports and
require that these reports are prepared
monthly by the Billing Department.

FINDING

8. There is no quality control function to ensure that:

* All users are adequately serviced. Account Represent-atives are assigned to some users, but the largest user, Paulis Register, has no assigned representative.

* All requests are processed on time.

* All requests are produced, verified, and sent out.

* Paulcomp's management receives daily reports of problems as well as status reports on critical production jobs.

9. Current pricing does not consider the use of premium software such as UFO. These products reduce develop-ment costs and time for the user but should be charged at a premium in production to compensate Paulcomp for the additional resources con-sumed, product cost, maintenance, and education.

CONCLUSION

A Quality Control/Data Audit function should be established to perform the following tasks:

* Verify input such as job control cards.

* Calculate batch totals.

* Verify output against requested output.

* Correct errors caused by inaccurate job control cards or batch totals and re-submit the job.

* Provide daily incidence reports to Paulcomp management.

* Allow weekly, monthly, and quarterly summary and analysis reporting.

Paulcomp should institute premium billing for the use of software products.

RECOMMENDATIONS

We recommend that a detailed analysis of the information
contained in this report be conducted with the objective
of implementing the conclusions we have reached.

The specific areas to be addressed include the current
pricing and billing problems, the installation of a cost
accounting system, the analysis and development of controls
for job accounting, and the definition of a quality assurance
function.

We further recommend that Jackson-Wilson Partners and the
management of Paulcomp review our conclusions, establish
priorities, and define target dates for implementation of
our conclusions. Once these steps are completed, Jackson-
Wilson Partners will develop a detailed plan to
implement the conclusions.

Exercises

1. Obtain a copy of a report on a campus-related issue or a copy of a corporate annual report. For purposes of classroom discussion—or written analysis—comment on the effectiveness of the report in terms of The PAFEO Process.

   For copies of reports on campus-related issues, consult your student government or newspaper.

   Copies of corporate annual reports are available in many libraries or directly from the corporations themselves free of charge. Many brokerage firms also distribute, free of charge, guides on how to read a corporate annual report.

2. Depending on your instructor's plan for this course, you may have to write a full-length formal report. Campus-related issues lend themselves to a number of assignments. Follow the directions of your instructor in generating partial or complete reports on the following issues.

   a. On-campus parking.
   b. Space assignments for student activities.
   c. Weekend or late-night access to the library, food services, or recreational activities.
   d. The justification for tuition increases.
   e. The effectiveness of admissions recruiting efforts.
   f. The effectiveness of campus public relations efforts.
   g. Campus security.
   h. Discipline in the dormitories.
   i. Access to campus computer terminals.
   j. Academic and other support services on campus.
   k. The writing problem(s) of students on your campus.
   l. A host of curricular issues.[2]

---

[2] The list originally appeared in Kevin J. Harty, "Campus Issues: A Source of Research in Business and Technical Writing Courses," *Teaching English in the Two-Year College*, 12 (October, 1985), 223.

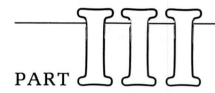

PART III

TOOLS

13

Words

In Parts I and II, we discussed writing for the world of work in terms of process and product. Here, in Part III, we will examine ways in which language choices affect (and are affected by) your purpose and your audience.

Words are the tools of the writer's trade. The ways in which a writer chooses words may be considered from several approaches.

Style is the result of the choices a writer makes in words, sentence structures, even in punctuation. These choices reflect the personality of the writer and the relationship he or she wishes to establish with the audience. A style may be judged by how well it achieves the writer's purpose and produces the desired response from the reader. As we shall see, in business there are styles which are hallowed by custom rather than by effectiveness. They are the products of misconceptions about the way business writing should work.

Grammar is concerned with correct words and sentence structures according to norms established by educated speakers and writers. Writers are expected by their readers to show a decent respect for "correct English," meaning that which agrees with the prescriptions of English handbooks.

Usage is concerned with appropriate words, and appropriateness is often a matter of taste. Recently, several dictionaries, such as *The American Heritage Dictionary* and *Harper's Dictionary of Contemporary Usage,* have asked panels of educated writers to comment on words or expressions as to their acceptability. The comments only serve to indicate how divided people are on what is acceptable usage. Where then does that leave the writer? Analysis of your audience can provide the only guidelines. You must try to be aware of the attitudes of your audience, remembering that the usage they approve is not always the one they regularly use.

In this chapter and the following one, we will try to share with you some advice on what to do and what to avoid if you want to cultivate habits of writing that lead to a clear, contemporary style and an acceptable grammatical usage.

Let's begin our discussion of style with a nod to a master of language, Sir Winston Churchill. In August, 1940, at the height of the Battle of Britain, Churchill wrote a memo to his staff—not about increased military preparedness, but about excess verbiage in interdepartmental correspondence:

> Let us have an end to such phrases as these: "It is also of importance to bear in mind the following considerations . . ." or "Consideration should be given to carrying into effect. . . ." Most of these woolly phrases are mere padding, which can be left out altogether or replaced by a single word. Let us not shrink from the short expressive word even if it is conversational.[1]

Evidently, the Battle of Britain was to be won on land, at sea, in the air, and with the typewriter.

In this chapter, we will talk about a number of problems you can get yourself into when you don't choose your words carefully. The words you choose can get in the way of your audience's ability to understand what you have to say. The words you use can even offend your audience, and an offended audience is even more difficult to communicate with. While there are a number of ways that you can go wrong with the words you choose, a little common sense and continued analysis of your audience can go a long way toward avoiding or solving problems that the words you choose can cause.

Stuffy and Out-of-Date Expressions

Because the world of business was once separate from the world of everyday life, writers in business and industry often felt that excessive formality had to be the hallmark of all letters, memos, and reports. Formality soon gave

---

[1] Quoted by staff writers for the Royal Bank of Canada in "The Practical Writer," *The Royal Bank Letter,* 62 (January/February, 1981), 3.

way to stiffness and stuffiness, and to an almost automatic use of stock phrases in every letter or memorandum.

Writers felt that the proper way to begin a letter was as follows:

> Per your letter of June 25, contained herewith is the information requested by you in same.

Unfortunately, today there are still enough people around who would not think twice about starting a letter the same way, or having their word processors start their letters that way—so much for automation! Of course, once you've started your letter with *per your* and *herewith,* you can easily continue in the same vein as the following letter sent out by an international shipping agent:

April 22, 1986
_____
(Date)

Harris/Costa Importers, Inc.
Five World Trade Center
Suite 798
New York, NY 10048

ATTENTION: Accounting Dept.   RE: M/S RIO DELGATO
                                       Invoice N14-4001
                                       Voyage 33 BHG
                                       Philapa 3/5/86

Dear Sirs:

We enclose herewith our duly supported disbursements accounting which is applicable to the above referenced.

We believe you will find each supporting invoice to be either self-explanatory or to have been previously explained. However, we respectfully solicit your inquiry regarding any item hereto attached about which you might have a query.

The enclosures reflect a balance of $ 16,890.87   due

~~our Company~~

your goodselves.

(✓) Our check in the above amount is affixed hereto.

( ) We would appreciate having your remittance in the
above amount at your early convenience.
 We remain

                                         Sincerely yours,

                                         Goldentree Bros.
                                         As Agents.

                                    BY: *E. H. Wilkins*
                                    _____

Enclosures

Often writers write such letters out of habit or because they have seen
others write this way. (We know of an insurance company that offers no
training to new employees who must answer queries or complaints from
insurance agents or policy holders. New employees are simply given a sev-
eral-inch thick book of sample letters to imitate. Some of the samples are
55 years old.)

Writing for business and industry requires clarity and conciseness. Excessive and meaningless formality undermine these traits.

Phrases to Avoid

While you can't write in the same way that you talk, you can check the
expressions you use in writing by asking yourself if you would ever use
those expression in talking face to face or over the telephone with someone.
You wouldn't walk into someone's office and say:

Per your letter of June 25, contained herewith is the information you requested
in the same.

Instead, you would say:

Here is the information you requested by letter on June 25.

You might even just say.

Here's the information you requested.

Write the same way. *Per your* and *herewith* are stuffy and out of date.
They hinder rather than help your reader. Avoid them, along with the following expressions:

pursuant to your
above referenced, above mentioned, above cited, above captioned
the above
below referenced, below mentioned, below cited

it has come to my attention
be assured, rest assured
be advised, kindly be advised
this is to inform you that
enclosed please find
herein, hereto
reference to same
thank you in advance, thanking you in advance
as of this date, as of this writing, as of the present
the undersigned
pending receipt of
in accordance with your request
we find that
we are of the opinion that
take under consideration, take under advisement
due to the fact that
to whom it may concern
under separate cover

Gobbledygook and Doubletalk

A second word-related problem you'll need to avoid is gobbledygook or doubletalk, the use of roundabout expressions to say simple things. For most of us

Illumination is Required to be Extinguished Upon Vacation of These Premises.

is much more simply put as

Turn Out the Lights When You Leave

Government has long been characterized as one of the major offenders in the use of gobbledygook and doublespeak, but no profession or field of endeavor is free from the problem.

At least one agency of the government, the Commerce Department, has tried to reverse this trend by reducing the gobble and resurrecting simple, straightforward prose. Commerce Secretary Malcolm Baldridge tells his employees to write "half way between Ernest Hemingway and Zane Grey with no bureaucratese."[2] Baldridge has even enlisted the aid of his department's word processors. Employees typing in words such as *viable, parameter,* or *prioritize* are told by their machines "Do Not Use This Word!" The

[2] On Secretary Baldridge's efforts, see *New York Times* (August 2, 1981), 33, and *Atlanta Constitution* (July 23, 1981), 1. We are grateful to Heidi B. Eddy, Special Assistant to Secretary Baldridge, for a copy of the Secretary's memo on writing style.
Other government agencies have also taken a second look at the writing their employees do. See, for instance, *Be a Better Writer, A Manual for EPA Employees* (Washington, DC: Environmental Protection Agency, 1980). Copies of this publication are available from the EPA Printing Management Office.

Secretary's approach will not cure all writing problems, but it may help readers wade through documents a little easier. A summary of his guidelines follows.

Secretary Baldridge banned the following as affected or imprecise words:

viable	target or targeted
input	effectuated
orient	output
hopefully (use *I hope*)	prioritize
ongoing (use *continuing*)	hereinafter
responsive	parameter (use *boundary* or *limit*)
specificity	image
utilize (use *use*)	inappropriate
thrust	optimize
maximize	finalize

He also had no use for the following phrases:

I share your concern (or interest or views)
I appreciate your concern (or interest or views)
I would hope (use *I hope*)
I regret I cannot be more responsive (or encouraging)
I am deeply concerned
Thank you for your letter expressing concern (use *Thank you for your letter regarding*)
prior to (use *before*)
subject matter
very much
bottom line
at this present time
as you know, as I am sure you know, as you are aware
more importantly
needless to say
it is my intention
mutually beneficial
contingent upon
management regime

The Secretary further cautioned against using redundancies such as:

serious crisis
personally reviewed
new initiatives
enclosed herewith
important essentials
final outcome
future plans
end result
great majority
untimely death

Finally, Baldridge told his staff not to use any of the following:

Please let me know if I can be of further assistance
I hope this information is helpful.
due to (use *because*)
self-initiate
overview (use *review*)
facilitate
I personally reviewed (use *I reviewed*)
dialogue
alternatives (use *choices*)
I understand that
I am pleased to designate
positive feedback
I or we believe
enhance
essence of
in terms of
point in time
time frame
unique

Gobbledygook when compounded turns into doubletalk or doublespeak. When faced with such doubletalk, your readers may wonder not only what you are trying to say but also what you are trying to cover up. The Watergate principals turned doublespeak into an art. No one involved in Watergate ever lied. They simply "misspoke themselves"—or perhaps their audiences "misheard" them. Jimmy Carter described his administration as an "incomplete success." The Gambling Commission in New Jersey called members of organized crime "members of career offender cartels." The president of a major automotive company referred to declining sales as periods of "negative economic growth." The Reagan White House referred to tax increases as "revenue enhancements."

When doublespeak and gobbledygook don't earn you the confusion or suspicion of your readers, they may earn you some dubious distinctions. The director of the catering division at British Rail attempted to explain why there had been no dining car on a passenger's train as follows:

Whilst I can readily appreciate your frustration at the loss of breakfast, since in the circumstances you describe it is unfortunately true that in many cases where a catering vehicle becomes defective and both stores and equipment need to be transferred into a replacement car, this can only be done during the train's journey.

It is not of course possible to make the transfer whilst vehicles are in the sidings and the intensity of coach working is such that the train sets are not available to put into a platform at other times to enable the transfer to be carried.

The director's efforts earned British Rail England's Plain English Award for 1981.[3] Few companies can afford such bad publicity.

Jargon and Technical Terminology

Jargon can be confusing or hard-to-understand language. Jargon can also be simply the specialized language appropriate to activities or groups. As long as experts in one field talk or write to other experts in the same field, their use of jargon is no problem. When experts in a field talk or write to laypeople or experts in a second field, the use of jargon can become a serious stumbling block for an audience.

The best monitors for the proper use of jargon are common sense and continued audience analysis. No one wants to go to a physician and be told that he or she has a "boo-boo on the knee." At the same time, a diagnosis of Paget's disease requires more explanation for the average patient than the fact that the disease is also known as *osteitis deformans*. The average patient needs a nontechnical definition of the disease, a list of symptoms, and information about diagnosis, treatment, and outlook.

Jargon can also become a problem for audiences when otherwise familiar terms have specialized meanings for given occupations or in different contexts. *Strike* means one thing in baseball, another in bowling, another in labor relations, and another in an exchange of punches. Your bank may return one of your checks with a notation that the check "needs to be signed by maker." Thanks to Joyce Kilmer, we know fools make poems and God makes trees, but who makes checks? You may know what *maker* means in this context, but how many people randomly stopped on a street corner would know? How many people receiving the bank's notification will know?

Familiar terms can also carry negative or confusing connotations in different contexts. The letter from the Internal Revenue Service used earlier in Chapter 4 was an announcement to the taxpayer that he or she was to be audited. However, the writer never used the word *audit,* since the word in the context of anything tax-related strikes fear into the hearts of taxpayers across the country. Bankers routinely deal with delinquent accounts, but no letter to a customer should refer to that customer or his or her account as "delinquent." *Dwindle* is a common word when used as a verb. In medical jargon, it has become a noun, usually plural, referring to advancing years leading to death from old age.[4] The Metropolitan Trunk Facilities Manager is a fairly high-level executive in the telephone company. For the average person on the street, the title might just as well refer to someone whose job involves the storage of luggage.

[3] See *New York Times* (December 8, 1981), B-21.
[4] Cited by William Safire in "Listening to What the Doctor Says," *New York Times Magazine* (November 9, 1980), 16, 18.

You'll never be able to eliminate jargon or technical terminology from your writing, but you can monitor its use. The most effective monitors are, as we have said, common sense and continued audience analysis. The following excerpt from a job vacancy notice for a serials cataloguer—that's library jargon for someone in charge of newspapers and magazines—in a university library probably makes no sense to you:

> Requires master's from ALA-accredited library school, along with 1–2 years' paraprofessional or professional cataloguing experience, preferably in serials cataloguing. Working knowledge of AACR2, OCLC cataloguing subsystems essential. Authority file background and experience with related serials functions in a research library preferred.

But it doesn't have to make sense to you, unless you have the necessary training and experience in library science and are thinking of applying for the job.

The expanding field of computer information services is in danger of overwhelming the layperson with its jargon. Indeed, fear of computers may in part spring from bewilderment over such simple terms (from an experienced programmer's point of view) as *hardware* and *software*. The New York Stock Exchange, mindful that not all investors have advanced degrees in business, publishes a brochure on the language of investment that carefully explains for the layperson such terms as *bear, bull, depreciation, liquity,* and *sinking fund.*[5] Most of us could have used some kind of glossary during the crisis at Three Mile Island in 1980. Just how much radiation is there in a *rem?*

Familiar Words and Explanations

When your audience is unfamiliar with the vocabulary you are using, you can solve their problems in several ways. First, you can, when appropriate, substitute a familiar term for a more technical, less familiar term. "Acute pharyngitis" sounds serious, unless your physician explains the diagnosis more familiarly as a "very sore throat." A reader may be confused as to what exactly "liabilities" are. But most readers will know what it means to owe money. If you are still a student, you may be unsure of the meaning of "matriculate"; all you ever did was "enroll." "Hardware" generally suggests tools, nuts, and bolts. "Computer equipment" should alert your reader to the proper context of your sentence.

With jargon, then, as with any issue involving word choice, use a nontechnical or familiar word instead of a specialized or less familiar word when the change won't affect your meaning. This advice applies in all situations to so-called 25-cent words as well. A headline on an article in the *New York Times* once announced that a Manhattan church was going to sell

[5] *Glossary: The Language of Investment* (New York: The New York Stock Exchange, 1978).

off some of its property in midtown and use the proceeds to support its "eleemosynary activities." *Time* once referred to a popular television game show host as "oleagenous." By using such words, both publications either left most of their readers confused or sent them running to their dictionaries. *Eleemosynary* and *oleagenous* are both fine words, especially for hard-to-solve crossword puzzles. *Charitable* and *ingratiating* get the point across to the average reader much more effectively.

English has a rich and varied vocabulary, but your purpose in writing should not be to improve your or your reader's vocabulary. Your purpose should be to communicate clearly and effectively.

Definitions, Synonyms, Comparisons, or Examples

You can also solve problems your readers have with specialized vocabulary by using definitions, synonyms, comparisons, or examples to make technical vocabulary more accessible to your readers. A definition can be explicit

> *Demaoization,* the process of undoing the excesses of the recent past in China, suffered a set back today.

or implied by your context:

> We're *bullish* on the market. We expect stock prices to rise rapidly for the next several months.

A notation to take "acetaminophen" instead of aspirin suddenly becomes clearer with the substitution of a well-known synonym: "Take Tylenol instead of aspirin." Likewise, an explanation of an individual retirement account that begins

> Individual Retirement Accounts (IRAs) combine a number of features familiar to our customers who have savings accounts and pension plans.

uses the familiar as a starting point for a discussion of the unfamiliar. Finally, examples provide a useful means for communicating technical information to any audience as long as you watch the pace of your explanation by way of examples:

> Two techniques used in 3-D movies have been demonstrated successfully on television. Both of them involve superimposition of two slightly different pictures—one for each eye—into the same space, and the use of special glasses to sort them out. The older of the two techniques is called anaglyph. The other is polarization. In anaglyph, two different-colored pictures are put on the screen—for example, an all-red picture for the left eye and an all-green one for the right. The viewer wears glasses with a green lens on the left and a red one on the right, each lens filtering out the picture in the eye. So the left eye sees only the red picture and the right eye only the green, in colors roughly resembling black and white.[6]

[6] David Lachenbruch, "Here Come 'Deepies'—Maybe," *TV Guide,* 29 (March 14, 1981), 23.

Legalese and Plain English

In 1978, New York became the first state to pass a Plain English Law requiring that consumer contracts be plainly written in everyday language. The first business to be sued under that law was the Lincoln Savings Bank in New York City. The offending document was a customer agreement about safe deposit boxes. The agreement contained the following sentence:

> The liability of the bank is expressly limited to the exercise of ordinary diligence and care to prevent the opening of the within-mentioned safe deposit box during the within-mentioned term, or any extension or renewal thereof, by any person other than the lessee or his duly authorized representative and failure to exercise such diligence and care shall not be inferable from any alleged loss, absence or disappearance of any of its contents, nor shall the bank be liable for permitting a co-lessee to have access to and remove the contents of said safe deposit box after the lessee's death or disability and before the bank has written knowledge of such death or disability.

This sentence even puzzled the state's Attorney General, Robert Abrams, who in bringing suit against the bank said, "I defy anyone, lawyer or layperson, to understand or explain what that [sentence] means."[7] The bank lost the suit and agreed to rewrite the original 121-word sentence as follows:

> Our liability with respect to property deposited in the box is limited to ordinary care by our employees in the performance of their duties in preventing the opening of the box during the term of the lease by anyone other than you, person authorized by you or persons authorized by the law.

The rewrite is still too long for one sentence, but at least all the legal mumbo jumbo is gone.

The movement to require plain English in all public documents is spreading. The impetus behind the movement has not, however, been simply a desire to cut the lengths of sentences and eliminate unnecessary legal terminology. The impetus is a desire to make consumer documents such as leases, warranties, and insurance policies understandable so customers can be better able to abide by their provisions.

Legalese, the special language seemingly intelligible only to lawyers, has several causes. Often consumer documents reflect policies spelled out earlier in city, state, and federal regulations. The regulations themselves are riddled with legalese, and writers of public documents find it easier to parrot or mimic rather than translate such regulations. There is also a perception that legalese insures legality. Yet wherever documents have been rewritten in plain English, those doing the rewriting have worked carefully with company lawyers. No suits have ever been filed on the basis of a document's having been translated from legalese into plain English.

[7] *New York Times* (August 31, 1980), 35.

Writing with the Consumer in Mind

The switch from legalese to plain English requires that a document not only drop all the *hereinunders* but also that a document change its basic philosophic approach. The document obscured by legalese seems designed to keep information from consumers. The document opened up by plain English is designed to inform the consumer.

You can translate a document from legalese into plain English in a number of ways. First, however, you need to pay careful attention to the law. Where important legal distinctions exist, you must maintain those distinctions. Where wording itself has been the subject of litigation, you can't change the wording. Otherwise, try using the following suggestions to open up consumer documents to consumers:

1. Reorganize the document to present information in a sequence and at a pace appropriate to your reader.
2. Use a personal tone by substituting forms of the pronouns *I* and *you* for impersonal nouns such as *Bank* and *Cardholders*.
3. Use simple language devoid of any legal trapping.
4. Use shorter sentences with active voice verbs.
5. Use contractions.
6. Use white space, headings, and lists to catch your reader's attention and to convince your reader that the document is accessbile to someone other than a lawyer.[8]

Here are some examples of legalese translated into plain English.[9] The first example comes from a personal loan form used by New York's First National City Bank (now Citibank):

OLD

In the event of default in payment on this or any other obligation or the performance or observance of any term or covenant contained herein or in any note or other contract or agreement evidencing or relating to any obligation or any Collateral on the Borrower's part to be performed or observed; or the undersigned Borrower shall die; or any of the undersigned become insolvent or make an assignment for the benefit of creditors; or a petition shall be filed by or against any of the undersigned under any provision of the Bankruptcy Act; or any money, securities or property of the undersigned now or hereafter on deposit with or in the possession or under the control of the Bank shall be attached or become subject to distraint proceedings or any order or process of any court.

---

[8] Alan Siegel, President of Siegel & Gale, is the pioneer in the business of simplifying contracts and consumer documents. See his two articles in *across the board:* "To Lift the Curse of Legalese—Simplify, Simplify," 14 (June, 1977), 64–70; and "The Plain English Revolution," 18 (February, 1981), 19–26.

[9] These three examples are quoted from *Time,* 106 (September 22, 1975), 74.

NEW

I'll be in default:

1. If I don't pay an installment on time; or
2. If any other creditor tries by legal process to take any money of mine in your possession.

The revision uses a list, contractions, and personal pronouns. More importantly, it leaves out a great deal of legal mumbo jumbo that the bank's lawyers obviously had second thoughts about including. This mumbo jumbo must either have been unnecessary or have been potentially problematic from a legal point of view.

Here's a second translation from legalese to plain English in a Sentry automobile insurance policy:

OLD

If the company revises this policy form with respect to policy provisions, endorsements or rules by which the insurance hereunder could be extended or broadened without additional premium charge, such insurance as is afforded hereunder shall be so extended or broadened effective immediately upon approval or acceptance of such revision during the policy period by the appropriate insurance supervisory authority.

NEW

We'll automatically give you the benefits of any extension or broadening of this policy if the change doesn't require additional premiums.

This revision also uses contractions and personal pronouns and eliminates unnecessary legal mumbo jumbo. The distinction between extension and broadening is, however, kept because it is a necessary and important legal distinction.

Here's a final example of translation from the Master Charge (now Master-Card) agreement issued by the First National Bank of Boston:

OLD

Cardholder and any other person applying for, using or signing the Card promise, jointly and severally, to pay to Bank the principal of all loans plus, as provided in paragraph 4, FINANCE CHARGES. Payments shall be made each month at Bank as Bank may direct, on or before the Payment Due Date, in the amount of (a) the greater of $10 or an amount equal to $\frac{1}{36}$th of the Total Debit Balance not in excess of the Maximum Credit on the related Statement Date plus (b) any amounts owing and delinquent plus (c) any excess of the Total Debit Balance over the Maximum Credit.

NEW

You must pay us a monthly minimum payment. This monthly minimum payment with be $\frac{1}{36}$ of the balance plus, of course, any amounts which are past due, but at least $10. If the balance is less than $10, the minimum payment will be the entire balance. The balance will include the outstanding amount that you have borrowed plus a finance charge.

Again, the revision uses personal pronouns. Here too the rewrite substitutes a series of plainly written shorter sentences for the original mammoth sentence and all its subsections.

Sexism

Sexism is discrimination based on gender. Sexism can be a problem in your writing when you compose messages that consciously or unconsciously slight members of either sex, though women have traditionally suffered more from sexist writing than men.

Here's a passage with a sexist bias:

> Every student should prepare his tentative roster before seeing his adviser. The student and his adviser can then discuss the roster in light of the student's academic needs and the career possibilities that are open to him. In addition, the student can, if he wishes, arrange to see a counselor from the Placement Office before he completes registration.

Unless the passage came from the regulations issued by the handful of remaining all-male colleges in this country, this passage is biased against women. It assumes all students are men.

The problem here lies with the pronouns. In English, we have different forms of the third person singular pronouns based on gender, and we lack an all-purpose third person generic singular pronoun other than *one. One* has, however, an unfortunate tendency to sound pompous.

Two solutions to the sexist bias in the passage just quoted that we *do not* recommend are (1) using *he/she* or *he* or *she* instead of *he,* and (2) explaining away the use of the masculine only as "traditional" in a footnote or an endnote. Extended use of the first of these two solutions in either form creates awkward and wordy sentences:

> Every student should prepare his/her tentative roster before seeing his/her adviser. The student and his/her adviser can then discuss the roster in light of the student's academic needs and the career possibilities that are open to him/her. In addition, the student can, if he/she wishes, arrange to see a counselor from the Placement Office before he/she completes registration.

> Every student should prepare his or her tentative roster before seeing his or her adviser. The student and his or her adviser can then discuss the roster in light of the student's academic needs and the career possibilities that are open to him or her. In addition, the student can, if he or she wishes, arrange to see a counselor from the Placement Office before he or she completes registration.

At best, the second solution, the explanatory note, pays lipservice to the attempt to be nonsexist when writing:

The traditional use of the masculine form of the personal pronoun in situations where the gender of the referent is unknown or irrelevant should not in any case be interpreted as conscious or unconscious sex discrimination in any section of this guide.

Unfortunately, it may be read as such discrimination and anger your audience.

A third solution, simply using *they* in place of *he/she* and *he or she,* causes grammatical problems:

Every student should prepare their tentative roster before seeing their adviser.

Their is plural and can't, therefore, refer back to the singular subject "student."

A fourth solution, invent a nonsexist generic singular pronoun, is innovative but flies in the face of the common sense we have been urging throughout this chapter. What would any reader make of the following version of the passage?

Every student should prepare nis tentative roster before seeing nis adviser. The student and nis adviser can then discuss the roster in light of the student's academic needs and the career possibilities that are open to nim. In addition, the student can, if nir wishes, arrange to see a counselor from the Placement Office before nir completes registration.

Nir, nis, nim don't run the risk of being sexist; they run the risk of startling or being unintelligible to the reader. Your goal in writing is not linguistic innovation. Your goal is effective communication, and you can meet that goal without being innovative or offensive.[10]

You can easily avoid sexism by capitalizing on the flexibility of the English language. Here are some problematic sentences and suggested revisions that eliminate sexism.

CHANGE
A lawyer will charge you an hourly rate for his services.

TO
A lawyer will charge you an hourly rate for legal services.

This revision simply eliminates the problem by eliminating the possessive pronoun and substituting an adjective.

You can also shift from singular to plural throughout your text.

CHANGE
A lawyer will charge you an hourly rate for his services.

[10] Ray A. Killian's *Managers Must Lead!* rev. ed. (New York: AMACON, 1979) uses a generic pronoun throughout to eliminate problems of sexism. Killian uses *hir* instead of *he/she* and *him/her; hirs,* instead of *his/her* and *his/hers; hirself,* instead of *himself/herself.* The resulting text is at times annoying to the reader. For a defense of the usage, see the publisher's note in Killian's book, pp. xv–xvii.

TO
Lawyers will charge you an hourly rate for their services.

You can also switch from third person to second person throughout your text.

CHANGE
A student's final grade will be based on his average score on his four examinations.

TO
Your final grade will be based on your average score on your four examinations.

Finally, you can sometimes substitute a definite or indefinite article for the possessive pronoun.

CHANGE
A student's final grade will be based on his average score on his four examinations.

TO
A student's final grade will be based on an average score on the four examinations.

These four solutions are nonsexist and sensible. Remember one monitor for choosing your words is common sense.

A second problem with sexism can arise with titles for jobs and positions that make unnecessary sexual distinctions, again often at the expense of women. Male and female writers alike are *authors* and *poets*. Thespians of either sex are simply *actors*. *Authoress, poetess*, and *actress* are sexist terms.

Television announcers used to tell us Joan Rivers was Johnny Carson's guest *host*, so the distinction between *host* and *hostess* is unnecessary in all contexts. *Waiter* and *waitress* do present a dilemma.[11] A Greenwich Village restaurant once advertised for a *waitron*. We have no idea who—or what—responded to the advertisement, but we would suggest simply advertising for "a waitress or waiter." *Table server* seems a title more appropriate to a *bus person*.

Where bravery is an issue, men and women can be equally heroic; therefore, Jill can be just as much a *hero* as Jack. Indeed, *heroine* has sexist and condescending connotations since the protoypical heroine is someone like Little Nell cruelly victimized by some monstrous male.

There is no real problem in distinguishing Congress*man* Tip O'Neill from Congress*woman* Shirley Chisolm when you know the sex of the person to whom you are referring. When you don't, *representative* or *member of Con-*

---

[11] On the odd chance that you ever have to write to or about royalty and nobility, English inherited the sexual distinctions enforced by *princess, countess*, and *duchess* from French in the twelfth century.

gress or *the House* presents an acceptable nonsexist alternative. Similarly, both *city councilman* or *alderman* and *city councilwoman* or *alderwoman* can be easily changed to *city council person* or *member*. Both *assemblyman* and *assemblywoman* can be changed to *assembly member* or *state representative*.

Your use of *chairman* needs to take note not only of possible sexism but also of personal preference on the part of the person in the chair who may prefer any of the following alternatives: *chair, chairman, chairwoman,* or *chairperson*. The person filling the chair's sex notwithstanding, possible nonsexist alternatives include *moderator* for those chairing meetings and *department head* for those chairing academic or other departments.

Academe presents three further problems: *alumnus,* professor *emeritus,* and *co-ed*. In borrowing the first two terms from Latin, English also had to borrow a grammatical distinction it doesn't share with Latin. In Latin, nouns have gender, so a distinction needs to be made between *alumnus* and *alumna* in the singular and between *alumni* and *alumnae* in the plural. *Alumns* and *alumpersons* will only startle—or amuse—your readers. Simply using *graduate* would, of course, easily avoid the issue of sexism.

Similarly, you need to distinguish a professor *emeritus* from a professor *emerita* and professors *emeriti* from professors *emeritae*. *Retired* professor may solve a number of problems and offers a commonsensical solution to a potential problem with sexism—or with the declensions of nouns in Latin.

Co-ed is sexist when used to refer only to women. Any student, male or female, attending a co-educational school is properly a co-ed.

When using and creating job titles, you should also be conscious of possible sexism. Some titles ending with *-man* can be simply changed to end with *-person*. Here are some further suggestions from the U.S. Department of Labor.[12]

Change	*To*
anchor man	anchor person or anchor
camera man	camera operator
coat/hat check girl	cloakroom attendant
craftsman	skilled worker (or a specific occupational title such as plumber or carpenter)
draftsman	drafter
fireman	fire fighter
foreman	supervisor or boss
insurance man	insurance agent or broker

---

[12] See *Job Title Revision to Eliminate Sex- and Age-referent Language from the Dictionary of Occupational Titles,* 3rd ed. (Washington, DC: U.S. Department of Labor, 1975). This publication is issued by the Office of *Man*power Administration!

Change	*To*
junior executive	executive trainee
lineman	line installer
longshoreman	stevedore
maid	houseworker or keeper
mailman	letter carrier or postal worker
male nurse	nurse
middleman	liaison or go-between
newsman	reporter
office boy or girl	office helper or assistant
policeman	police officer
salesman	salesperson or clerk
spokesman	spokesperson, speaker, or representative
statesman	leader, diplomat, or public figure
weatherman	weather forecaster or meteorologist
workman	worker

Throughout your writing, you should also provide men and women with parallel treatment. When you refer to men as "men," you can't refer to women as "girls." If you wouldn't describe a man by his appearance, don't describe a woman by her appearance. If you refer to men simply by their last names, do the same for women.

Finally, you'll want to avoid the danger of sexism in the salutations and addresses you use in correspondence. Again, common sense and audience analysis can guide you. If you know that your reader is male or female or married, unmarried, or liberated, you'll run into no problems when you use the courtesy titles *Mr., Mrs., Miss,* or *Ms.*

When you don't know your reader's sex or marital status, don't run the risk of offending that reader. H. R. Jones could be a man or a woman. If a woman, she could be single or married and object to or not mind being called *Ms.* If you can't find out your reader's sex or marital status, drop the courtesy title completely. Here's an inside address and salutation without such a title:

```
H. R. Jones
Vice President
Factory Outlets, Inc.
P.O. Box 92
Providence, RI 02907

Dear H. R. Jones:
```

Some women dislike being referred to as *Ms.* If you know Jones is a woman but can't find out her marital status or her personal views on the use of *Ms.,* again drop the courtesy title:

```
Hanna Royce Jones
Vice President
Factory Outlets, Inc.
P.O. Box 92
Providence, RI 02907
```

```
Dear Hanna Royce Jones:
```

You'll also need to avoid the danger of sexism when you write simply to a company or business address and you don't have the name of an individual and can't find one out. Common sense should tell you to pick up the telephone and ask. But if you really can't find out or decide you really don't need to use an individual's name, your address can simply read:

```
Factory Outlets, Inc.
P.O. Box 92
Providence, RI 02907
```

But what saluation would you use here? The following are all sexist. They presuppose the reader is male:

```
Dear Sir(s):
Sir(s):
Gentlemen:
```

"Ladies and Gentlemen" seems more appropriate for speeches. "Sir or Madam" sounds a bit stiff, and how many women appreciate being called "madam"? "To whom it may concern" also suffers from stiffness, and a fair number of people have problems with *who* and *whom,* so they will write "To who it may concern." "Gentlepersons" or "gentlepeople" is silly, as is "Dear Factory Outlets" or "Dear Company." Simply eliminate the saluation here. It is unnecessary; the only other options are sexist or silly.

If you have no name, but you do have an attention line, you may be able to get a salutation out of that attention line:

```
Factory Outlets, Inc.
P.O. Box 92
Providence, RI 02907
```

```
ATTENTION: Personnel Director
```

```
Dear Personnel Director:
```

Again, though, use common sense.

The Administrative Management Society suggests using the simplified letter format we discussed in Chapter 4 as yet another way of avoiding sexism in business letters. This format replaces the saluation with a subject line typed in all capital letters. Not only does this line serve as a possible solution to problems of sexism, but it also introduces the reader to the purpose of the letter and serves as a handy reference and filing guide.

Words and Phrases That Are Easy to Confuse

The following letter exemplifies a final problem you can have with the words you use in your letters, memoranda, and reports. The letter was, by the way, actually sent out by a bank.

```
Dear Sir:

We have discovered a mystic in your account. Please
contact our office immediately to strengthen out your
account. We have been unable to reach you by telephone.
```

The problem here is with the words *mystic* and *strengthen,* which the writer has somehow confused with *mistake* and *straighten.* While real, the example is admittedly extreme. The problem—many words and phrases sound alike or have similar meanings—is, however, all too real for many writers. The following list differentiates some of the more common problematic words and phrases. When in doubt, reach for your dictionary.

Accept/Except: *Accept* is a verb. It means "consent to," "admit to," or "agree to take." *Except* is a preposition. It means "excluding" or "other than."

    Mr. Smith *accepted* the package from the delivery person.
    All employees, *except* those who plan to retire next year, should meet in the dining room.

Adapt/Adept/Adopt: *Adapt* and *adopt* are both verbs. *Adept* is an adjective. *Adapt* means "adjust to new circumstances." *Adopt* means "take as one's own." *Adept* means "skilled."

    We will have to *adapt* our marketing campaign to our new market.
    Janet Ritz is an *adept* administrator who has helped us out on a number of occasions.
    The company *adopted* my suggestion that we stagger work hours for the duration of the transit strike.

Advice/Advise: *Advice* is the noun. *Advise* is the verb. *Advice* means "counsel." *Advise* means "give counsel" or "suggest."

    My *advice* was that John should not apply for the position until he had completed training.
    I *advised* John not to apply for the position until he had completed training.

Affect/Effect: *Affect* can only be used as a verb. It means "influence." *Effect* as a noun means "result." *Effect* as a verb means "cause" or "bring about."

The President's conduct *affects* the public's opinion of all elected officials.
The *effects* of your physical fitness program are noticeable.
Once in office, the Mayor *effected* a number of positive changes in the tax laws.

Agree to/Agree with: If you *agree to* something, you are "giving consent." If you *agree with* something, you are "in accord" with it.

I *agree to* pay you back in weekly installments of $25.00.
I *agree with* your suggestion that we stagger work hours during the transit strike.

All Right/All-Right/Alright: *All-right* and *alright* are not words. *All right* is a phrase meaning "all correct." It does not mean "acceptable" or "good."

Your answers on the object examination were *all right.*

Change: Is your pay *all right?*
  To: Is your pay *acceptable?*

All Together/Altogether: *All together* is a phrase meaning "acting in unison" or "completely in one place." *Altogether* is a word meaning "completely" or "totally."

We finally got the litigants *all together* in the judge's chambers.
Your comments were *altogether* uncalled for.

Allude/Refer/Elude: *Allude* means "making an indirect reference to." *Refer* means "making a direct reference to." *Elude* means "escape detection or notice."

Stop *alluding* to the problem. Explain it directly.
He made a point of *referring* to my client three times by name in his testimony.
The burglar *eluded* the police officer who had chased her for three blocks.

Allusion/Illusion: An *illusion* is a false image or misperception. An *allusion* is a veiled or indirect reference.

My boss is under the *illusion* that I enjoy working here.
The careful reader will notice many *allusions* throughout the report to happy times in the company's past.

Alot/A lot/Allot: *Alot* is not a word. *A lot* is an informal phrase, generally inappropriate for writing done in business and industry. The phrase means "many" or "a great deal." *Allot* is a verb meaning "distribute," "apportion," or "allocate."

I *allotted* the speaker ten minutes for her presentation.
The delegates were *allotted* a number of first-class hotel rooms.

Change: I took *a lot* of time to finish the proposal.
  To: I took *a great deal* of time to finish the proposal.
Change: *A lot* of people helped with the report.
  To: *Many* people helped with the report.

Already/All ready: The adverb *already* is used to express time. The phrase *all ready* means "completely prepared."

The shipment has *already* been sent.
I'm *all ready* to start work on Tuesday.

Among see **Between**

Amount/Number: *Amount* is used for things thought of in bulk. *Number* is used for things that can be counted.

> The *amount* of work required for the project was underestimated.
> The *number* of people required to complete the project was underestimated.

Appraise/Apprise: *Appraise* means "estimate the value, significance, worth, or status." *Apprise* mean "inform."

> The jeweler *appraised* the worth of the watch.
> The engineer *apprised* the vice-president about the status of the repair work.

As see **Like**

Awhile/A while: The adverb *awhile* means "for a short time." *A while* is a phrase meaning "a period of time."

> The doctor thinks we can wait *awhile* before surgery.
> He waited for quite *a* long *while* before he called the police.

Beside/Besides: *Beside* means "next to." *Besides* means "other than" or "in addition to."

> I sat *beside* my sister at the head table.
> *Besides* the two of us, the head table also contained my parents.

Between/Among: *Between* relates two items or persons. *Among* relates three or more items or persons.

> I sat *between* Bill and Dave at the dinner.
> Unfortunately, the three of us were not *among* friends.

Biannual/Biennial: *Biannual* means "twice during the year." *Biennial* means "every second year."

> Our *biannual* meetings were held in January and June.
> The thirteenth *biennial* meeting comes just two years to the day of the strife-ridden twelfth meeting.

Capital/Capitol: *Capital* refers either to financial assets or to a city that is home to a legislative body. *Capitol* refers to the actual building in which a legislative body meets.

> Baton Rouge, the *capital* of Louisiana, is home to the tallest *capitol* building in the United States.
> We'll need twice as much *capital* before we can make a downpayment on that house.

Cite/Site/Sight: *Cite* means to "quote an authority." A *site* is a plot of land or a place. The ability to see is *sight.*

> Professor Jones *cited* Adam Smith's controversial theories repeatedly.
> Political considerations will dictate the *site* for the new convention center.
> John lost his *sight* in a work-related accident and sued his employer for $3 million.

Compare/Contrast: When you *compare,* you point out similarities. When you *contrast,* you point out differences.

> I *compared* the estimates, and I found no real differences.

The Sawyer sample *contrasts* with the Burger sample in that the Sawyer is reliable and affordable.

Complement/Compliment: A *complement* completes something else. *Compliment* means "praise." Either word can be used as a noun or a verb.

A *complement* of three more officers is needed to fill out the honor guard.
The speaking course *complements* the writing course.
The buyer's *compliment* could mean a big raise for you.
Miss Jones *complimented* Mr. Davis on his fine performance.

Continual/Continuous: *Continual* means "frequently repeated." *Continuous* means "unbroken."

His *continual* complaints about the quality of the food in the cafeteria finally did some good.
The noise of the jackhammer was *continuous* from dawn to dusk.

Contrast see **Compare**

Council/Counsel: A *council* is a group that discusses, advises, or consults. *Counsel* is the advice such a group might give. *Counsel* as a verb refers to the giving of such advice. Lawyers both are and offer legal *counsel*.

City *Council* meets every Thursday.
Both the accounting department and the legal staff offered the same *counsel:* sell the property.
The chairwoman *counseled* against making rash judgments.

Credible/Creditable: Anything *credible* is believable. Anything *creditable* should be given praise or credit.

Your presentation was *credible;* the statistics made your point dramatically.
You did a *creditable* job in your presentation; the boss wants to give you a bonus as a result.

Data/Datum: *Data* is the plural of *datum.* This distinction notwithstanding, there is considerable confusion and disagreement over what form to use with *data.* Some sources argue that *data* has become a singular collective noun in much the same way that *group* has. We lean toward retaining *data* as a plural.

These *data* are correct. This *datum* is correct.
The *data* are incomplete; therefore, they cannot be trusted.

One way to avoid the problem is to substitute another word for *data.*

These *facts* are correct. This *fact* is correct.
The *findings* are incomplete; therefore, they cannot be trusted.

Diagnosis/Prognosis: A *diagnosis* is an analysis of the nature of something. It can also be the result of such an analysis. A *prognosis* is a prediction.

The *diagnosis* was cancer, but the *prognosis* for recovery was excellent since the disease had been caught in an early stage.

Differ From/Differ With: Things *differ from* each other. *People differ* with each other.

My background *differs from* yours a great deal.
We *differ with* each other about the best solution to the problem.

Different From/Different Than: When what follows *different* is a clause, use *different than.* When what follows *different* is a phrase, use *different from.*

My house is *different from* yours and Bill's. Mine has a two-car garage.
The estimate was *different than* we had expected.

Discreet/Discrete: You are *discreet* when you show tact. *Discrete* means "separate" or "distinct."

I am always *discreet* about personal information contained in the files.
The *discrete* units on the order form are color-coded.

Disinterested/Uninterested: *Disinterested* parties are fair and objective. *Uninterested* parties don't care. They have "no interest" in what's at hand.

We need a *disinterested* person to settle this dispute.
Despite my repeated suggestions, the auditor remained *uninterested* in the advantages of the new calculator.

Economic/Economical: *Economic* is the adjective form of *economics. Economical* means "not wasteful."

The *economic* recovery was stalled by rising gasoline prices.
Because we have a limited budget, we need an *economical* solution to our membership problems.

Effect see **Affect**

Elude see **Allude**

Eminent/Imminent: Someone who is an expert or an authority is *eminent.* Something which is about to happen is *imminent.*

Einstein was an *eminent* scientist.
The end of the strike was *imminent.*

Except see **Accept**

Explicit/Implicit: An *explicit* statement is clearly expressed. An *implicit* statement is only implied.

You were *explicitly* told not to park in the company lot.
The President talked about our problems with inflation. *Implicitly,* he blamed those problems on his opponents, though he never referred to them by name.

Fewer/Less: *Fewer* items can be counted. *Less* refers only to mass amounts.

There are *fewer* paydays this month, and I expect to take home *less* pay as a result.

Foreword/Forward: *Foreword* is a noun referring to an opening statement or section, usually in a book. *Forward* is an adjective or adverb meaning "at or toward the front."

I have been asked to write the *foreword* to his new book.
The *forward* luggage compartment is full. Please store your belongings in the rear of the airplane.
To accelerate the mechanism, turn the dial *forward.*

Healthful/Healthy: *Healthful* things promote good health. *Healthy* things have good health.

Wheat germ is *healthful*.

He is remarkably *healthy* for a man his age.

Illusion see **Allusion**

Imminent see **Eminent**

Implicit see **Explicit**

Imply/Infer: Writers and speakers *imply* by hinting at or suggesting something. Readers and listeners *infer* by drawing their own conclusions.

The chairman *implied* that profits were lower than had been expected.

After hearing the chairman's presentation, the stockholders *inferred* that the company didn't expect to grow much in the next several years.

Insoluble/Unsolvable: *Insoluble* means "incapable of being dissolved." *Unsolvable* "incapable of being solved."

The material is *insoluble* even in boiling water.

The production problem is evidently *unsolvable*.

Irregardless/Regardless: *Irregardless* is unacceptable as a substitute for *regardless*, a word meaning "no matter what" or "unmindful."

Change: I will not comply with your request *irregardless* of the law.

To: I will not comply with your request *regardless* of the law.

Its/It's: *Its* is the possessive of *it*. *Its* means "of it." *It's* is a contraction of *it is*.

It's important that the delivery arrive by Tuesday.

The company lost *its* option on the property.

Lay/Lie: *Lay* is a transitive verb meaning "place" or "put." *Lay* requires a direct object. The principal parts of *lay* are *lay, laid,* and *have laid*. *Lie* is an intransitive verb meaning "remain" or "recline." *Lie* does not require a direct object. The principal parts of *lie* are *lie, lay,* and *have lain*.

He *laid* the carpet on Monday before I arrived with the new furniture.

The doctor told the injured women to *lie* still.

Leave/Let: As a noun, *leave* means "a period of time." As a verb, *leave* means "depart." *Let* is a verb meaning "allow" or "permit."

My medical *leave* of absence begins on Monday.

He *left* for the coast yesterday.

Let the chips fall where they may.

Less see **Fewer**

Libel/Slander/Liable/Likely: To *libel* someone, you must injure that person's reputation in print. To *slander* someone, you must injure that person's reputation in speech.

After reading your profile of your opponent in the paper, I am surprised you weren't sued for *libel*.

I heard your speech, and I intend to sue you for *slander*.

Liable means "responsible for." *Likely* means "probably."

Because the injury occurred after normal work hours, the company is not *liable* for damages.

It is *likely* that I will move to the suburbs sometime after the first of the year.

Like/As: *Like* is a preposition. It can be followed by a noun or a pronoun. *As* is a conjunction and is used before clauses.

> He did *as* I told him to.
> He is just *like* his mother.

Loose/Lose: *Loose* is an adjective meaning "unfastened" or "unrestrained." *Lose* is a verb meaning "misplace" or "fail to win."

> The *loose* wire shorted out the entire floor.
> If we *lose* the account, I'll be fired.

Maybe/May Be: Both mean "perhaps." *Maybe* is an adverb. *May be* is part of a verb phrase.

> *Maybe* I will go home early today.
> I *may be* buying that new car after all.

Media/Medium: *Media* is plural. *Medium* is singular.

> The effect of television as a *medium* for influencing social change is underestimated.
> Representatives of all three *media*—radio, television, and the newspapers—gathered in the auditorium.

Notable/Noticeable: *Notable* means "worthy of notice." *Noticeable* means "readily or easily observed."

> Her contributions were *notable* in a number of areas.
> The dent is hardly *noticeable*.

Number see **Amount**

Oral/Verbal: *Oral* means "spoken." *Verbal* means simply "in words" and could mean *oral* or *written*.

> There was no contract; we simply had an *oral* agreement.
> To avoid possible legal problems, I'll draw up a *verbal* agreement in the morning.

Passed/Past: *Passed* is a verb, the past tense of "pass." *Past* is a noun referring to a period in time.

> I *passed* Joan Kenner in the hall yesterday.
> He didn't graduate because he didn't *pass* two of his courses.
> In the *past*, we have made too many costly mistakes.

Personal/Personnel: *Personal* means "relating to a person." *Personnel* refers to a group of people usually sharing common employment.

> Skip the section asking your age and other *personal* questions.
> All military *personnel* are banned from this area.

Phenomenon/Phenomena: *Phenomenon* is singular. *Phenomena* is plural.

> The latest *phenomemon* is glow-in-the-dark hoola hoops.
> The economy suffered because of a number of natural *phenomena*, especially the great blizzard of last March.

Principal/Principle: *Principal* refers to an amount of money or a chief officer. *Principle* refers to a truth or belief.

> *Principal* and interest owed amount to $3.9 million.
> When I was in first grade, I spent half of every school day sitting in the *principal's* office.

The fundamental *principle* here is whether the federal government has a right to interfere with our purchasing the stock.

Prognosis see Diagnosis

Raise/Rise: Both *raise* and *rise* mean "move to a higher position." *Raise* always takes an object because it is a transitive verb. *Rise* is an intransitive verb and never takes any object.

The bond market is expected to *rise* steadily for the next month.
We informed our customers that we would *raise* our prices 10 percent starting in June.

Refer see Allude

Regardless see Irregardless

Set/Sit: *Sit* is always an intransitive verb indicating position. The principal parts of *sit* are *sit, sat,* and *have sat. Set* can be transitive or intransitive. *Set* means "put," "place," "harden," or "establish."

Let's *set* a date for our meeting now.
Once the glue *sets,* you can drink out of the glass.
Janet *set* the file on the shelf after she *sat* down.
Why don't you *sit* here while I get you some coffee?

Sight see Cite

Site see Cite

Slander see Libel

Some Time/Sometime/Sometimes: *Some time* refers to a period of time. *Sometime* refers to an unknown period of time. *Sometimes* refers to repeated occurrences.

I waited until *some time* after you called to see Bill.
We will have to meet *sometime* after Christmas.
Sometimes Bill forgets to pick up his paycheck.

Stationary/Stationery: *Stationary* is an adjective meaning "immobile" or "unchanging." *Stationery* is writing paper and envelopes, or, more generally, any material used for writing or typing.

The copy machine is *stationary;* it is too heavy to move between offices.
The next time we order *stationery* we must remember to get twice as many large mailing envelopes.

That/Which/Who: *That* and *which* refer to animals and things. *Who* refers to persons.

The files *that* you borrowed are needed today.
The store, *which* opened last year, burned to the ground.
The teacher *who* wins the prize will receive $5,000.00.

See also **Who/Whom.**

There/Their/They're: *There* is an adverb indicating place. *Their* is the possessive of *they. They're* is a contraction of *they are.*

I went *there* often when I was younger.
The students all forgot *their* umbrellas.

They're three reasons not to fill your return early.
There were six different Democratic candidate in the primary.

To/Too/Two: *To* is used as a preposition. It is also part of an infinitive. *Too* means "excessively" or "also." *Two* is a number.

I went *to* Shreveport.
It is wrong *to* misuse company property.
We paid *too* much for this meal.
I agreed to quit *too.*
Two typewriters were stolen.

Uninterested see **Disinterested**

Unsolvable see **Insoluble**

Verbal see **Oral**

Which see **That**

Who/Whom: *Who* is used in place of *he, she,* or *they. Who* is a subject. *Whom* is used in place of *him, her,* or *them. Whom* is an object.

Who is the dean of the college?
She is the dean.
Whom did you vote for?
I voted for *them.*

Choosing between *who* and *whom* in sentences that make statements rather than ask questions is more difficult. To determine which word to use, look at the function of the word in the sentence. *Who* functions as a subject. *Whom* functions as an object.

The policeman *who* will drive the car is tall.
The policewoman *whom* they shot recovered from her wounds.

See also **That.**

Who's/Whose: *Who's* is a contraction of "who is" or "who has." *Whose* is the possessive form of "who."

Who's responsible for the cost overrun?
Whose cost overrun are we talking about?

Your/You're: *Your* is the possessive form of "you." *You're* is a contraction of "you are."

Your company lost money last quarter.
You're not being considered for the promotion anymore.

Exercises

1. An exclusive men's clothing story notifies new customers that their applications for charge accounts have been approved with the following engraved notice:

```
Dear Mr. Burke:

We gratefully acknowledge the receipt of your recent
application for credit and with pleasure advise of the
approval of your account.

We shall be glad to have you avail yourself of its
privileges and assure you of our appreciation for the
opportunity of serving your requirements.

                            Faithfully yours.

                            Men's Clothiers, Inc.
```

What do you and your classmates think of this letter?

2. You are familiar with the story of Little Red Riding Hood. Discuss the "improvements" Russell Baker has made in the following revision (*New York Times Magazine* [January 13, 1980], 10).[13]

Little Red Riding Hood Revisited

In an effort to make the classics accessible to contemporary readers, I am translating them into the modern American language. Here is the translation of "Little Red Riding Hood":

Once upon a point in time, a small person named Little Red Riding Hood initiated plans for the preparation, delivery and transportation of foodstuffs to her grandmother, a senior citizen residing at a place of residence in a forest of indeterminate dimension.

In the process of implementing this program, her incursion into the forest was in midtransportation process when it attained interface with an alleged perpetrator. This individual, a wolf, made inquiry as to the whereabouts of Little Red Riding Hood's goal as well as inferring that he was desirous of ascertaining the contents of Little Red Riding Hood's foodstuffs basket, and all that.

"It would be inappropriate to lie to me," the wolf said, displaying his huge jaw capability. Sensing that he was a mass of repressed hostility intertwined with acute alienation, she indicated.

"I see you indicating," the wolf said, "but what I don't see is whatever it is you're indicating at, you dig?"

Little Red Riding Hood indicated more fully, making one thing perfectly clear—to wit, that it was to her grandmother's residence and with a consignment of foodstuffs that her mission consisted of taking her to and with.

At this point in time the wolf moderated his rhetoric and proceeded to grandmother's residence. The elderly person was then subjected to the disadvantages of total consumption and transferred to residence in the perpetrator's stomach.

[13] Copyright © 1980 by the New York Times Company. Reprinted by permission.

"That will raise the old woman's consciousness," the wolf said to himself. He was not a bad wolf, but only a victim of an oppressive society, a society that not only denied wolves' rights, but actually boasted of its capacity for keeping the wolf from the door. An interior malaise made itself manifest inside the wolf.

"Is that the national malaise I sense within my digestive tract?" wondered the wolf. "Or is it the old person seeking to retaliate for her consumption by telling wolf jokes to my duodenum?" It was time to make a judgment. The time was now, the hour had struck, the body lupine cried out for decision. The wolf was up to the challenge. He took two stomach powders right away and got into bed.

The wolf had adopted the abdominal-distress recovery posture when Little Red Riding Hood achieved his presence.

"Grandmother," she said, "your ocular implements are of an extraordinary order of magnitude."

"The purpose of this enlarged viewing capability," said the wolf, "is to enable your image to register a more precise impression upon my sight systems."

"In reference to your ears," said Little Red Riding Hood, "it is noted with the deepest respect that far from being underprivileged, their elongation and enlargement appear to qualify you for unparalleled distinction."

"I hear you loud and clear, kid," said the wolf, "but what about these new choppers?"

"If it is not inappropriate," said Little Red Riding Hood, "it might be observed that with your new miracle masticating products you may even be able to chew taffy again."

This observation was followed by the adoption of an aggressive posture on the part of the wolf and the assertion that it was also possible for him, due to the high efficiency ratio of his jaw, to consume little persons, plus, as he stated, his firm determination to do so at once without delay and with all due process and propriety, notwithstanding the fact that the ingestion of one entire grandmother had already provided twice his daily recommended cholesterol intake.

There ensued flight by Little Red Riding Hood accompanied by pursuit in respect to the wolf and a subsequent intervention on the part of a third party, heretofore unnoted in the record.

Due to the firmness of the intervention, the wolf's stomach underwent ax-assisted aperture with the result that Red Riding Hood's grandmother was enabled to be removed with only minor discomfort.

The wolf's indigestion was immediately alleviated with such effectiveness that he signed a contract with the intervening third party to perform with grandmother in a television commercial demonstrating the swiftness of this dramatic relief for stomach discontent.

"I'm going to be on television," cried grandmother.

And they all joined her happily in crying, "What a phenomena!"

3. Translate the following into plain or at least easier-to-understand English.[14]

[14] The source for exercises a and b is Debra Shore, "'Identify Me by the Nomenclature of Ishmael . . .'," *Brown Alumni Monthly,* 81 (February, 1981), 20–23. The source for the remaining exercises is "Government, Business Try Plain English for a Change," *U.S. News & World Report,* 83 (November 7, 1977), 46.

a. From a university course announcement:

Tuition regulations currently in effect provide that payment of the annual tuition entitles an undergraduate degree candidate to full-time enrollment, which is defined as registration for three, four, or five courses per semester. This means that at no time may an undergraduate student's official registration for courses drop below three without a dean's permission for part-time status and that at no time may the official course registration exceed five.

b. From the back of an Amtrak ticket:

Times shown on timetables or elsewhere and times quoted are not guaranteed and form no part of this contract. Time schedules and equipment are subject to change without notice. Amtrak expressly reserves the right to, without notice, substitute alternate means of transportation, and to alter or omit stopping places shown on ticket or timetable. Amtrak assumes no responsibility for inconvenience, expense, or other loss, damage, or injury resulting from error in schedules, delayed trains, failure to make connections, shortage of equipment, or other operating deficiences.

c. From a promissory note:

No extension of time for payment, or delay in enforcement hereof, nor any renewal of this note, with or without notice, shall operate as a waiver of any rights hereunder or release the obligation of any maker, guarantor, endorser or any other accommodation party.

d. From an insurance policy:

The insured may pay the amount of ultimate net loss to the claimant to effect settlement and, upon submission of due proof thereof, the company shall indemnify the insured for that part of such payment which is in excess of the underlying limits, or, the company will, upon request of the insured, make such payment to the claimant on behalf of the insured.

e. From the regulations governing citizens' band radios:

The current authorization, or a clearly legible photocopy thereof, for each station (including units of a station) operated at a fixed location shall be posted at a conspicuous place at the principal fixed location from which such station is controlled.

4. Rewrite the following memo correcting any problems in word choice.

March 1,1983

TO: All Department Chairmen
FROM: Fred Lizzio, Controller
RE: 1983-84 Budgets, Allocations for Supplies

    As you all know, it is now time to prioritize funds for the 1983-84 budgets within the parameters of growth specified by the Board of Trustees. Needless to say, the tasks that lay ahead will be difficult. Budget cuts will effect all areas of our institution. I am, however, confident that the amount of cuts we will have to make will not disrupt any one department all together.

At the same time, we cannot work this coming year under any allusions. Across the board cuts will and must be made. Working with your deans, I have all ready made certain preliminary cost savings. This memorandum simply lists the next steps to be taken in the area of supply disbursement and allocation.

Beginning with the month of July, departmental allocations and disbursements for supplies will be cut in the amount of 8%. The principal behind this across the board cut is simple. Money for supplies is easier to cut then money for other items in the budget. Their is also a feeling that considerable waist occurs due to careless use of stationary and other supplies. It is impossible to determine specifically whose to blame for such waist, but unnecessary and unauthorized personnel use of supplies has at least been a factor in the matter at hand in the passed. Lose inventory control maybe a second factor here.

Due to the fact that it is more economic to centralize inventory control through my office, henceforth all such control will be so centralized, where uninterested members of my staff can monitor acceptable levels of use. At such times when difficulties arise, some sort of compromise among individual chairman and I can and will be worked out.

If we work together in monitoring expenditures for the abovementioned budget item, I am sure no chairman will find his department suffers unnecessarily. I hope I can count on your cooperation. If I can be of assistance to you at any time, please don't hesitate to call or see me.

CHAPTER 14

Getting It Right: Grammar, Punctuation, and Spelling

When you take off your writer's hat, it's time to don your editor's green eyeshade and play a different role. As editor of your own work, you are no longer the creative phrase-maker; you are now the keen-eyed, cold-blooded editor looking for problems that may distract or confuse your audience. The most obvious distractions and confusions result from errors in grammar, punctuation, and spelling.

How, you may ask, do such mechanical matters interfere with communication? Because they rudely violate readers' expectations, mechanical or grammatical errors break the fragile thread that connects writer to reader. Readers stop thinking about your message and start thinking about you and your manners. Yes, manners. Whether they are conscious of it or not, readers are accustomed to the considerate good manners of the magazines and newspapers they read. They are used to reading Standard English that is well-punctuated and free of spelling errors. Violating these expectations of Standard English is a social as well as a writing error. It's poor manners—like eating your peas with a knife or slurping your soup. It annoys the reader, who expected better of you. Your reader begins to wonder whether you are not a dunce after all, and your credibility is damaged.

Why chance it? Study this chapter carefully. If you learn what is here, there is a good chance you'll be able to eliminate the most frequent and most annoying errors in grammar, punctuation, and spelling.

Common Grammatical Problems

Agreement

Everyone knows the basic rule: subjects and verbs agree in number, and pronouns agree with their antecedents. What then is the problem?

Actually, there are two problems. The first is to identify the real subject and not be misled by intervening phrases and clauses. The second lies in determining whether a particular subject is in fact singular or plural, a decision that is especially tricky when you have an indefinite pronoun or a collective noun as a subject. These examples illustrate the problems.

Discussion of demographics, promotional costs, and marketing strategies *have* been going on for the past six months.

Since the real subject of the sentence is *discussion*, the verb should be singular, *has*. Confusion can arise because the intervening nouns closer to the verb (*demographics*, *costs*, and *strategies*) are all plural. But what has been going on is a discussion, not demographics, costs, or strategies. The subject is, therefore, *discussion*.

Neither management nor the striking unions *cares* about the public.

The verb should be plural, *care*, to agree with the part of the compound subject nearest to it, *unions*.

Some of the files were destroyed, but I can assure you that some of that money rightfully belongs to me.

Indefinite pronouns such as *some, none, all, more,* and *most* may take either the singular or plural, depending on whether the writer wishes to communicate a sense of a single amount or a mass (as in *money* in the example above) or a plural number of countable units (as in *files* in the same example).

The contents of the files *was* a well-kept secret.
The contents of the boxes *were* divided equally.

When the meaning encompasses the group taken as a whole, the collective noun takes a singular verb. When the meaning suggests a collection of individual items or units, the plural verb is used.

Words like *each, every, anyone, anybody, no one, nobody, everyone,* and *everybody* ordinarily take singular verbs. The problem in agreement sometimes arises when a pronoun later in the sentence refers to this singular antecedent. How does this sentence sound to you?

Everybody was supposed to hand in their assignments today.

That's what most people would say in informal English, isn't it? But in formal writing they would be expected to write

Everybody was supposed to hand in his or her assignment today.

But let's see what happens if we apply the same rule of formal agreement in another example.

Everyone came to the meeting, and the chairman welcomed *him.*

You can see that the consistent use of the singular here is formally correct but that it is ridiculous in view of the meaning intended. In cases like this one, the meaning takes precedence over the form. When the indefinite pronoun is singular in meaning, it is referred to by a singular pronoun; when plural in meaning, it is referred to by a plural pronoun. In the example above, the pronoun should be *them.*

(In the previous chapter, we discussed the agreement of pronoun and antecedent to examine some of the possible problems in sexism that may arise because the English language lacks a singular pronoun to refer to an antecedent of mixed or indefinite gender.)

Pronoun Reference

A pronoun is a word that replaces a noun. The word it replaces is called its *antecedent.* If we didn't have pronouns, we would have to write silly sentences like "John put John's hat on John's head." Because they're such handy little devils, we take pronouns for granted in our everyday speech. In doing so, we form some bad habits for writing.

We omit antecedents.

Someone was stealing from the company, but *they* couldn't prove it.

We ignore ambiguous references.

John asked Kevin if *he* could help out.

We use *which, this,* or *that* with broad reference to a general idea rather than to a specific noun or pronoun antecedent.

The boss was curious about his assistant's sudden affluence. *This* led him to make some discreet inquiries.

All of these examples would probably be overlooked in informal speech; all of them could be sources of confusion in writing.

To avoid problems with pronoun references in your writing, keep this principle in mind: *The noun to which a pronoun refers should be unmistak-*

able. When you edit your own work, look suspiciously at every pronoun. Ambiguity is your enemy. Root it out. Just because you know what you mean is no assurance that your reader will. Try to write so the reader *cannot* mistake your meaning.

Parallel Structure

Do you see anything wrong with these sentences?

1. The representative's job is to visit clients, discuss their needs, and then he should report to the home office.
2. Common sense does not govern either the formation of company policy or how it is carried out.
3. She was clever but a dishonest employee.

These sentences are as smooth as the sound of a fingernail running down a blackboard. They grate on the ear and the mind, because they fail to put related thoughts in a parallel grammatical structure that clarifies the logical relationship of the ideas.

Sentence 1 contains a series of three elements that ought to be parallel verb for verb—*visit, discuss,* and *report*. Instead, the last element breaks the parallel and disrupts the reader's expectations. The sentence should be revised to clarify the parallel ideas.

The representative's job is to visit clients, discuss their needs, and report to the home office.

Sentence 2 demonstrates the need for parallelism with correlative pairs like *either/or, neither/nor, both/and,* and *not only/but (also)*. The grammatical structure on one side of the correlative must be repeated on the other side of it. Since a noun (*formation*) follows *either*, a noun must follow the *or*.

Common sense does not govern either the formation of company policy or the implementation of it.

The conjunctions *and, but, or, nor,* and *yet* should generally signal a match of noun for noun, adjective for adjective, dependent clause for dependent clause, and so on. Sentence 3 should, therefore, be revised to read

She was clever but dishonest.

Right about now a little voice deep within you is probably saying, "Does anybody really care about parallel structure except English teachers?" The answer may very well be no—at least on the conscious level of analysis. Your reader is unlikely to say, "Aha! A violation of parallel structure!" Nevertheless, without quite knowing why, the reader has a sense of the grating fingernail on the blackboard. Something seems irritatingly wrong. The idea is not coming across as clearly and efficiently as it should.

Clarity and efficiency are the twin rewards of maintaining parallel structure. They result from fitting the syntax (the arrangement of words) to the meaning (the arrangement of ideas). Have you ever driven a car with one bad spark plug? It lacks both power and performance, and it wastes gasoline sinfully. Replace the plug and you have a glorious change in power and efficiency. Parallel structure has a similar effect on your sentences. Why inflict a bumpy ride on the reader?

Restrictive and Nonrestrictive Clauses

Understanding the differences between these two kinds of clauses will help you to clarify for yourself and your readers the relationship between the main idea of the sentence and the secondary idea. Without this understanding, you will have problems with punctuating—problems that will cause confusion for your reader. It is, therefore, worth taking a few minutes now to get the distinction between restrictive and nonrestrictive clauses straight.

A *restrictive clause* provides information about the main clause that is essential to the meaning the writer intends. (The "*that* clause" in the previous sentence is a good example.) A clause is restrictive when you find that you can omit the relative pronoun (*who, which,* or *that*). The restrictive clause should *not* be set off by commas before and after it.

> The plan *that the president rejected* would have cost the company ten million dollars. (The restrictive clause identifies the plan you are talking about and is, therefore, essential to your meaning.)

A *nonrestrictive clause* adds useful information to the sentence, but it could be dropped without blurring or changing the main idea. To signal the reader that it is parenthetical information, you set it off by inserting commas at the beginning and end of the clause.

> The plan for a merger, which the president rejected, would have cost the company ten million dollars.

When the relative pronoun could be either *which* or *that,* it is a good general practice to use *that* for restrictive clauses and *which* for nonrestrictive clauses.

Punctuation

Despite the impression you may have, punctuation does not consist of endless lists of rules known only to English teachers and others of that ilk. More often, punctuation is a matter of common sense and taste, combined with a knowledge of certain well-established conventions.

Let's begin with a few general principles you can apply when you are not sure about a specific case.

1. Punctuation exists to help the reader follow your thought. Use punctuation to prevent misreadings, to meet the demands of established conventions, and to clarify the sense units of a sentence.
2. Don't use a punctuation mark you can't justify on the basis of one of the three purposes above. Overpunctuating causes more confusion than underpunctuating.

The Period

A period at the end of a group of words tells the reader that you have expressed a complete thought. Every complete thought should have a period at the end, unless it has a question mark or an exclamation mark.

You can do two things wrong with periods, and both of them are serious barriers to communication.

1. You can put them at the end of a group of words that don't express a complete thought and thus create the confusion known as an illegitimate *sentence fragment.*

> The new product will not be available until after Labor Day. Since it will take that long to complete the required consumer testing.

The second thought must be attached to the first one if it is to make sense. Its dependency is signaled by the word *since.* If you have difficulty identifying fragments, be suspicious of word groups that begin with *such as, like, who, which, when, where, before, after, since, because, although, while, unless, until.* These words usually indicate that the word group needs to be attached to the clause preceding or the clause following them. The other serious error is a sin of omission.

2. You can omit the period after a complete thought and allow it to run on into the next thought.

This is a *run-on sentence,* a serious error that forces the reader to stop to puzzle out your meaning.

> Our costs have gone up dramatically our labor rates have risen sharply.
> We anticipate adding new services to handle them we will also require new staff.

The Comma

The comma is probably the most widely used, most useful mark of punctuation. Because it has so many uses, writers and editors often disagree on usage. Where does that leave you?

Often you find yourself floundering between "Thou shalt" and "Thou shalt not," depending on which handbook you consult. Actually, the situation isn't all that bad. Although there are gray areas of disagreement about comma usage, you can take a giant step towards mastery if you learn a few widely used conventions. Here are some you ought to put in your memory bank:

Dates. Use commas after the day and the year. If no day is mentioned, no commas are needed.

> I took a step that changed my life on August 7, 1954.
> I took a step that changed my life in August 1954.

Addresses. Use a comma after each element in the address *including the last.*

> Send replies to my office at 16 Lakeview Drive, Watkins Glen, New York, and I will give them my immediate attention. (The comma after *New York* is often left out and shouldn't be.)

Numerals. Use commas to separate thousands.

> His personal debts totaled over $1,200,000.

Quotations. Commas and periods go *inside* the quotation marks—always, even when it does not seem to make sense. This convention is a typographical convenience for printers, not a grammatical law. In Great Britain, the convention is to put the marks outside the quotation marks. You should follow the standard U. S. practice: commas and periods go inside; semicolons and colons go outside.

> He called his column "Marginalia," a title once used by Edgar Allan Poe.
> "The cost," he explained, "is not the only thing that concerns me."
> The chapter was entitled "Getting It Right"; it dealt with grammar, punctuation, and spelling.

When you have made these conventional uses habitual, you will have eliminated many of the problems of comma usage. If you learn only four more guidelines, you will have sufficient mastery of the comma to use it confidently and well. Here are the four guidelines:

1. *Compound Sentences:* Put a comma before the conjunction (*and, but, or, nor, for*). It clarifies the structure of the sentence for the reader and prevents any possible misreading.

> I saw the progress report for May, and the one for June does not promise to be much more encouraging.

2. *Series:* Put a comma before the final *and* when you are separating items in a series. By doing so, you'll avoid any possible confusion.

> At our table there was a man who ate no meat, a woman who ate only bread, and children who ate only the dessert. (Could be startling without the comma: imagine a woman who ate only bread and children.)

3. *Introductory Phrases and Clauses:* Use a comma after a long introductory phrase or clause (one that exceeds six words). If the opening phrase or clause is short but likely to cause misreading, use the comma as a safeguard.

> Among those who know him best, he is considered a highly talented amateur chef.
> If he wishes, the money can be paid to his trust fund. (Could be easily misread without the comma.)

4. *Parenthetical and Nonrestrictive Elements:* These are phrases or clauses that could be enclosed in parentheses because the sentence would be complete and coherent without them. They require a comma *before* and *after* to set them off from the main idea of the sentence.

The mayor, who ran on an economy platform, now has announced the biggest tax rise in the city's history.

Ms. Gloria Donnelly, Director of the Nursing Department and a nationally recognized authority on stress management, is the keynote speaker.

The Semicolon

Many excellent business writers and journalists get along fine without ever using a semicolon. You can too, and it would be better to banish the semicolon from your writing than to use it indiscriminately, as some students do. But why discard a useful tool, one that is useful in signaling subtle relationships between the parts of a sentence?

It is not difficult to learn the correct uses of the semicolon if you remember that it can function either as a strong comma (a supercomma, if you will) or as a weak period (a semiperiod).

As a *supercomma,* it can be used to separate clauses that already have internal commas.

The scholar seeks to know the past; the report writer is concerned with the present, generally with a view to plotting the future; hence, their outlooks must be quite different.—Jacques Barzun

Note how the semicolons neatly divide the principal parts of this lengthy sentence, making it easier for the reader to see the way the writer has structured the relationships.

Since shorter sentences are more characteristic of business writing, however, the semicolon is more frequently used as a *semiperiod,* a stop sign rather than a red light. It tells readers to stop at the end of one independent idea, but it also allows them to roll slowly through to the next idea, which is closely related.

To use the semicolon correctly as a weak period, remember one thing: use it only between independent clauses, those capable of standing alone as sentences. Never use it between a dependent clause and a main clause. If you are using it correctly as a semiperiod, you should be able to substitute a period for it.

RIGHT
The idea was not entirely his own; to be truthful, it was entirely his wife's.

WRONG
While other firms have recognized the importance of the new technology and put it to use in their marketing strategies; we continue to act as though nothing has really changed.

Why is the second example wrong? If you're not sure, it would be a good idea to go back and read this section on the semicolon again.

One final reminder. Don't confuse the semicolon with the colon. It should never be used after the salutation of a business letter.

RIGHT
Dear Dr. Mall:

WRONG
Dear Dr. Mall;

The Apostrophe

Even the authorities don't always agree on the rules for the apostrophe. In general usage, it appears wrongly in advertising, on signs, on labels, on mailboxes, on greeting cards. If you don't want to join the ranks of assassins by apostrophe, here are two things to avoid and three simple guidelines to follow.

1. Don't use the apostrophe to show possession if the word is already a possessive pronoun like *hers, its, ours, yours, theirs, whose.*

    The company is having its logo redesigned. (not *it's,* which means *it is*)

2. Don't use the apostrophe to indicate the plural.

    The mailbox should say *The Fallons,* not *The Fallon's.*

Here are minimal guidelines which should carry you through most difficulties:

1. To make a noun possessive, simply add the *'s* (*editor's desk, boss's orders, children's toys*).
2. If the noun already ends in *s* because it is plural, add only the apostrophe to make it possessive (*students' rights, players' association, secretaries' hours*).
3. If the word is a proper noun (the name of a particular person or place), add *'s* to form the possessive (*Dickens's life, James's novels, Marx's writings*).

Besides indicating the possessive of nouns, the apostrophe also signals a contraction by filling the space where a letter or numeral is omitted. (*don't, doesn't, the summer of '83*)

The Hyphen

The best advice we can give regarding the hyphen is to get a good desk dictionary such as *Webster's Ninth New Collegiate* and follow the recommended hyphenation for the word in question.

In general you use the hyphen

- When you're using two or more words as a unit to describe a noun (*fast-buck artist*).
- When you have to divide a word at the end of a line (check the dictionary).

A Final Word on Punctuation

Obviously, these few pages won't answer every problem you can possibly encounter regarding punctuation. We have chosen to be carefully selective, dealing only with those problems writers encounter most frequently. At the risk of incompleteness, we chose practicality and manageability. It is possible to learn these few pages well and thus be prepared for most situations you encounter in everyday writing. A complete handbook covering virtually every possibility, on the other hand, seems to us to defeat its own purpose by intimidating the student. "Are you kidding? I'll never remember all those rules!"

So what do you do when you encounter a problem not covered here? Everyone who has to write job-related assignments ought to have a few reference books on hand. These would include a good desk dictionary, a handbook of English, and a style manual. These few references can provide answers to any brain teasers you might come across, such as: "Do I capitalize the names of seasons?" (No), or "When do I use brackets?" Incidentally, if you're wondering why we did not include capitalization in this section, we believe that most matters of capitalization are matters of common sense; the ones that aren't are matters of choosing a consistent style and sticking to it. That's why it is handy to have one style manual on your desk that you can follow consistently.

In summary, we remind you that the basic purpose of punctuation is to link or separate elements within sentences so as to make the meaning clear to your reader. While you will find many variations in punctuation usage among skilled writers, you need not be confused by differences. Use all the punctuation required by current conventions, as outlined here, and use only as much optional punctuation as necessary to help your reader follow your meaning. Avoid unnecessary punctuation, which slows the reader and may even blur your meaning by causing pauses in inappropriate places.

Spelling

If you learned everything about writing contained in this book and still made spelling errors frequently, you would be judged a poor writer. That's how much spelling matters. Spelling errors can lower your grades on papers and examinations and make you seem less intelligent than you really are. Spelling errors alone can disqualify job applications, cost your company business, or prevent your promotion. Several misspellings on a page produce a "reverse halo" effect that puts you and your entire message in a poor light. Are you now convinced that spelling really matters?

Since it matters so much, you must do everything in your power to eliminate misspellings from your written work. Unfortunately, there is no easy shortcut to becoming a better speller. But you take the first important step toward becoming a better speller the moment you recognize how important correct spelling is to your readers and set out to eliminate misspellings from your own work.

There isn't any single way to learn to spell. Memory is certainly important, but what kind of memory? Some good spellers have eye memory: they look at a word closely and remember how it is supposed to look on the page. Others have ear memory: they sound out the word and remember the syllable sounds. Still others have muscle memory: they remember a word better if they actually write it down. Since it is uncertain what kind of memory works best for you, why not use as many of them as you can to help you become a good speller?

To become a better speller you have to begin by doing two things:

1. Work on those words that usually give you trouble, using as many different kinds of memory as you can.
2. Learn to be a careful proofreader by looking at each word separately and then at each syllable separately. Use the dictionary if you have the slightest doubt about a spelling.

When you have put together a list of the words that give you trouble, you want to work on getting the correct spelling of each of these words into your memory. To do that, you must apply eye, ear, and muscle memory to the task. Here's how.

- Stare at the word. Look for the precise part that causes you trouble. (Is it the *i* in hypocri*sy*? The *rr* and *ss* in emba*rr*a*ss*ment?) Use eye memory, and focus on the letter or syllable that confuses you.
- Sound it out. Exaggerate the syllable that is troublesome. (Make your ear hear that it's irresist*ible,* not irresista ble.)
- Write it out at least once or, better, three times. Let your hand muscles or your typing fingers get the feel of it.
- Make up a mnemonic, a memory jogger, to help you associate the correct spelling with something. (The princi*pal* was never my *pal.*) Associate prep*a*ration with prep*a*re, compe*ti*tion with compe*te,* defin*i*tely with defin*i*tion.

In addition to working on the words you don't know how to spell, you must also develop careful proofreading habits to catch careless or hasty mistakes in words you do know how to spell. Studies have shown that 85 percent of the errors in students' papers are the result of careless writing or inadequate proofreading rather than lack of knowledge.

Perhaps this percentage will drop dramatically when more people write on word processors with spelling programs to help with the proofreading. Until word processors are as common as typewriters, however, good proof-reading is still one of the writer's prime responsibilities. We have one strong suggestion: proofread at least once for spelling alone. Don't worry about punctuation or choice of words in this reading; concentrate on words and syllables and try not to think about the sense. Put a check in the margin next to any line containing a questionable spelling, and then check the dictionary for each of these words. When you find the correct spelling, apply the memory techniques just suggested to fix the word in your mind.

Your mind may stagger at the notion that you must learn to spell only by memorization. Although memory is terribly important because English spellings are often so illogical, it is possible and convenient to learn a few principles, each of which can be applied to many similar words. Let's look at five of the most common problems and try to formulate a principle which can be carried over whenever we encounter words that fit one of the five categories.

1. UNSTRESSED VOWELS   Unstressed vowels are a problem because the sound is so indistinct in pronunciation that the hearer cannot tell which vowel the sound represents. Is it gramm*a*r or gramm*e*r? Remember, the problem is only one letter. You can help yourself by accentuating that one troublesome vowel. Be conscious of this vowel when you pronounce the word. Write the word several times, capitalizing the unstressed vowel each time. In brief, stress the unstressed vowel.

2. *ie* OR *ei* WORDS   Words with *ie* in them usually rhyme with *niece; ei* words usually rhyme with *eight*. One of the most useful little verses ever written is that memorable poem that goes like this:

Write *i* before *e*
Except after *c*
Or when sounded like *a*
As in *neighbor* and *weigh*.

That bit of verse will carry you through words like *receive* and *conceit,* but it is not a panacea; there are exceptions to the rule, and there is nothing to do but memorize them. They include:

caffeine	height
codeine	heir
counterfeit	their
either	leisure
fiery	protein
financier	seize
foreign	weird

3. DOUBLING FINAL CONSONANTS   When adding a suffix like *ed* or *ing* or *able,* do you double the final consonant or not? The unhelpful answer must be, "Sometimes you do; sometimes you don't." There is a rule that can help you if you will take the pains to learn it.

- For one-syllable words ending in a single consonant following a single vowel (like *plan*), double the consonant before adding a suffix beginning with a vowel (like *able, ed, ing,* or *y*).
- For words of more than one syllable ending in one vowel and one consonant, double the final consonant if the word is accented on the last syllable. (This rule helps out with many problem words like *prefer, occur, transfer,* and *control.*)

Remember, we didn't say the rule was easy; we said it was helpful. Admittedly, some will find it easier to memorize a list of words that double the final consonant than to learn this rule.

4. WORDS THAT SOUND OR LOOK ALIKE One of the most serious spelling blunders is to confuse words that sound or look alike. Such a blunder can lead to confusion of meaning. ("The daze seemed endless.")

Since look-alike or sound-alike words account for some of the most serious spelling blunders, the ones that make readers slap their foreheads and say, "Oh my God, look at this!" you owe it to yourself and your readers to guard against such mistakes. Refer to the list in Chapter 13. Then find any words that give you trouble and work on them, visualizing which word you would write for each meaning. Often these errors result from vocabulary limitations rather than misspellings; the writer writes *discreet* to mean separate and distinct because he is unaware that there is such a word as *discrete*. To eliminate such errors, you must study meanings as well as spellings so you can match the proper spelling to the meaning you want.

If you follow the suggestions in this chapter, you *can* become a better speller. But it will take work on your part. Since misspellings obscure your purpose and distract or offend your audience, you must take care to eliminate them from your writing. If you are not a good speller, find someone who is and ask that person to proofread your work and circle the misspellings. Then consult your dictionary and correct the words yourself; don't have someone else correct your spelling. When you correct the word, use your eye, ear, and muscle memory to fix the correct spelling in your mind. If you want to be excepted as a writter who's work has the write affect on a reader, you have to no that mispellings can reek havoc on communication. (And if you didn't find eight errors in that sentence, you have just proved how much you need this section of the book!)

Exercises

1. Examine the following sentences closely for mistakes in grammar, punctuation, or spelling. Correct any errors you discover.

   a. Within the branch, AIS also has new support systems, that should work to their advantage in terms of productivity.
   b. The client with over 40 telecommunications users or more complex requirements demand reclassification so they can be handled by the Large Business Market.
   c. The personalizing function tailors the station to the individual user enabling each user to create and change distribution lists; telephone numbers where the user may be reached; office and home hours when user can be called, and placing on hold the automatic delivery of messages.
   d. I would appreciate your vote, as well as others you might speak to, in the May 17 primary.

e. Effective immediately, all properties located in the city of Philadelphia, on which this company holds a mortgage, will now be serviced from this location.

f. Each of the employees know their responsibilities.

g. Neither the president nor the division heads has recognized the effect of the new law.

h. His assistant told everyone she met about the mistake he had made. This cost her a raise.

i. Howard immediately called Jack because he owed him a favor.

j. Judy has a tendancy to view the problem as a personal one. This leads to unecessary conflict.

k. The company had become overly dependant on goverment subsidies.

l. The idea of a forty-eight hour week did not thrill the union representative.

m. There is a typewriter, a dictating machine and a word processor in your new office.

n. Before a worker retires, they should cultivate other interests outside of the job.

o. The most recent order, together with the rest of the letters from Havishams, have been misfiled.

p. If any one of the proposals are accepted, the company would be quite pleased.

q. He is the only one of the applicants who have not filed the required health report.

r. I decided to revise my resume. This took many additional hours but proved to be worth it.

s. The manager supported my position, which made me certain my proposal would be successful.

t. Everyone in the group agree on the most important issue.

u. Each customer is requested to take their used trays, dishware, and utensils to the nearest designated cart.

v. Series of Three Writing Workshops Prove Successful (headline from a newsletter of a Writers' Conference)

w. We are extremely proud of our city's heritage, which dates back to colonial days, as well as the many and varied attractions it has to offer.

x. If your credit card is ever lost or stolen, please report it to your nearest Bloomingdale's store immediately.

y. We'll do our best to help you quickly, courteously, and to your satisfaction throughout the entire contract period.

2. The following statement was written by a politician seeking the office of mayor in a large Eastern city. The question posed by the League of Women Voters was: "What specific action would you take to increase jobs in (this city)?"

If you were this candidate's press representative, how would you correct the punctuation and the grammar to make the statement more presentable?

First we must look at the financial base of the city, the inner city neighborhoods, the vacant factory buildings, the small business section of the city can be a start, we must first upgrade our educational curriculum in our public schools, and to provide adequate retraining for those out of school. I will appoint a peoples task force to help me, not corporation executives or academic research.

3. The following response to an information request demonstrates the chaos one can create with careless sentences and faulty punctuation. Rewrite it to correct the errors. (If you have completed Chapter 7, your instructor may want you to rewrite the letter completely to make it a more effective response to a query.)

Jan 5th 1987

Tucker Kinney
87 Lowell Drive
Framingham, MA 01701

Dear Tucker:

Thanks for your letter to our mail order dept. They sent it on down to our main showroom to answer your questions. While both the Rao & Thornton are good tables. And we do handle the Thornton.

We would suggest the CJ-Doran 55 turntable as the much better value. As this table is directly compared to the DAK LP-12 and such tables. The reveiws that have been written so far indicate that the Doran table is every bit as good as the Dak. And much better than the other tables in the same price range.

When the same arm and cartridge was tested on both the Dak and Doran the opinion was that the listeners could not tell the tables apart. Even when some of the listeners used their own records. With this in mind and at a savings of about $600.00 between tables. You would not have to think to hard as to the table of choice. Especially when the Doran does look as good or better than the Dak.

We also handle the Polhemus, Ludlow, English, Oldsey, and several other line of good cartridges. So if you need further information you might like to drop over or give us a call and we would be glad to answer any of your other questions. In the mean time thanks again for your letter. And best regards.

Sincerely,

J. Caufield Maxwell

J. Caufield Maxwell, Pres.

4. Using what you have learned in this chapter, revise the following short letter. Discuss the errors found with other members of the class.

Dear Dick:

The inclosed report represents the results of our initial survey of pricing practices at Welshcorp. As a result of this effort, Consultants Inc. is confident that we can assist you in establishing a a more meaningful pricing methodology. One that will equitably reward Welshcorp for your efforts, and will assure you and your mangement team, as well as your clients, of excellent service and fair prices.

Consultants Inc. proposes to initiate the detail analysis and functional design of this system immediately, in accordance with our inclosed recommendations.

We appreciate this opportunity to be of assistance to you, and we look forward to working with you on this critical project.

Very Truly Yours,

Ida Mae Wabash

Ida Mae Wabash
Consultant

Word Processing and the Writing Process

If you are not now writing on a word processor, the odds are that you will be before very long. Before the turn of the century, the electric typewriter will join the quill pen in the museum. That's what the predictions say, anyway.

Of course, we all know a good many writers who would rather die than give up their pencils and yellow legal pads. Many, who never could compose on a typewriter, will certainly never hand over their golden words to the erratic electronic monster called the computer. If you are sure you are one of these writers, you might want to skip this chapter (or burn this book!).

If, on the other hand, you are already acquainted with writing on a word processor, you will certainly read the chapter as an exercise in skipping. But most of our readers, we suspect, will fall into a middle group—curious, interested, but not too experienced in writing on a computer. Can it actually help me write, you wonder, or is it just a better way of transcribing what I have written—an electronic typewriter attached to a TV screen? Although word processing started out as a technological advance on the typewriter, it's now evolving into a tool that may help in the writing process itself, not just a tool for transcribing what has already been written.

Much research is now going on in an attempt to discover more about the effect of this new tool on thinking and writing. What follows is a *writer's* introduction to word processing. It assumes no knowledge of word processing on the reader's part. It's not a how-to-do-it manual. It explores the possibilities word processing has for the writer, not only in speeding up and easing the transcribing of words, but also in influencing each stage of the writing process.

What Is Word Processing?

It really isn't processing at all. You don't feed words into the machine and have them come out as finished pieces of writing. The idea of "processed words" is as unappealing to the writer as "processed veal" is to a gourmet. The term *processed* somehow connotes a sacrificing of substance to appearance. A word processor doesn't take control away from the writer. On the contrary, it allows more freedom to edit and revise. The writer types words, but the words do not go directly to paper. Instead, they are stored on a disk and projected on a screen. With a few key strokes, the writer can correct errors, insert or delete words or whole passages, move blocks of text around, check spelling, and establish or change the format in which the piece will appear on the page. Because the text exists only as electronic blips, it does not harden and take shape until the writer is ready. It appears on the monitor as a tentative arrangement, subject to instantaneous change or deletion by the writer. At a keystroke, the words move and the text rearranges itself once more; the screen reflects the dynamics of the writing process. Only when the writer is ready does the continual writing process stop and result in the product we call the printed page, which emerges at high speed from a printer driven by the computer. The word processor and the food processor do very different things, but they both save time and effort for the writer or chef, thus helping the creative person to do the best job possible.

As long as we're examining the meaning of the term *word processing,* we might as well clear up another source of confusion. The manuscript of this book was written (and processed) on a Zenith 150 microcomputer, not a word processor at all. This desktop computer became a word processor with the insertion of a diskette containing a word-processing program. There are machines called "dedicated" word processors, but they cost thousands of dollars and are generally used in large offices by trained operators skilled in rapid transcription of words, not in composing with words.

The microcomputers on which most writers work consist of a slightly expanded typewriter *keyboard* for entering letters, figures, or commands; a *computer,* which has one or two disk drives to hold the floppy disks that store the entered information; and a *monitor* or TV screen that allows the user to see an electronically printed version of the words written. Because microcomputers have limited memory in which to store what has been written, the words produced are stored electronically on a floppy disk, which looks something like a 45 rpm record. What is stored can be speedily printed on paper if the writer has a printer attached to the computer. A piece of writing, called a "document" in computer jargon, may be changed in wording or arrangement or format as much as the writer wishes before being printed on paper ("hard copy" in computerese). Since the words are still stored on the disk after printing, they may be further changed if the writer doesn't like the printed version.

The computer can't do any of these things until it is told how by a word-processing program. This program is fed into the computer's memory from a floppy disk. The computer is like a turntable without a record until it gets its program from the disk. The disks are called "software," to distinguish them from the "hardware" like the computer, keyboard, monitor, and printer. Word-processing software must be compatible with the computer. A program written for an IBM computer will not run on an Apple computer, for example.

What Can a Word-Processing Program Do?

According to *Research in Word Processing Newsletter,*[1] here are some of the capabilities a typical writer would need and be able to get from good word-processing software.

HELP SCREENS   These are instructions that can be called on to the screen whenever you need them. If you are not working daily with the program, you forget some of the commands you don't use often. The HELP key comes immediately to the rescue. Not all HELP screens are equally helpful, however. Be sure that the one you buy is clear enough to be useful.

AUTOMATIC HEADERS (TITLES), FOOTERS, AND PAGE NUMBERS AND FULL SCREEN CURSOR CONTROL   This feature allows you to move the little blinker called the cursor to any spot on the screen that requires editing, insertion, or deletion.

AUTOMATIC WORD WRAP   You don't have to press RETURN at the end of the line; the computer automatically goes to the beginning of the next line. This feature greatly increases typing speed.

ADJUSTABLE LEFT AND RIGHT MARGINS   This feature is needed if you write term papers and want to indent and single space a long quotation.

[1] "The Least You Should Know About Word-Processing Software," *Research in Word Processing Newsletter,* 2 (April, 1984), 6–7.

AUTOMATIC TEXT ADJUSTMENT AFTER INSERTING OR DELETING   It's magical the first time you see it. Delete a word or line and the words on the screen pirouette into place so quickly you hardly realize that your change has been incorporated into the clean text before you.

SINGLE- AND DOUBLE-SPACING AND CAPACITY TO SHOW FORMATTED TEXT ON SCREEN BEFORE PRINTING   This feature saves time and paper. It is especially important because you usually don't see a whole page on your monitor; most hold only 25 lines or less. For a writer, inability to view a page at a time while writing is somewhat of a handicap. Many programs claim to show the page exactly as it will print, but because of differences in the size of the screen and the size of the page, they do not. In some programs, you cannot make corrections while viewing the formatted text—an annoyance, but not a serious one, since you can return to the Edit mode and then correct the errors (if you still remember them and can find them!)

SEARCH AND REPLACE   If you misspell a word you have used more than once, you can have the computer search for that word and replace it with the correct spelling. The computer will search for any "string" you specify exactly, whether it be a character, a word, or a phrase. It will find (and replace if you wish) any deadwood expression like "types of" or jargon like "impact upon." The uses of the "Search and Replace" feature in editing your own work depend on how imaginative you are in making it work for you.

CUT AND PASTE   This allows you to move blocks of text from one place to another without any retyping. You can move a sentence or even a whole paragraph to reorganize what you have written.

Depending on the kind of writing you do, you may want to look for some other features too. For research papers, your program ought to be able to do superscripting for placing footnote numbers above the line. If you do any technical or scientific writing, you will want to be able to subscript the numerals in formulas. And many business writers will want a program that can do some graphics (like line graphs, bar graphs, and pie charts). Graphics capabilities are usually included in separate spread-sheet programs rather than in word-processing programs.

The price of word-processing software varies more than the number of features. In addition to the features we have described, you will want to look for readable manuals and good support from the software manufacturer. Prices for software containing all or most of the features listed above range from $50.00 to $500.00! Before you invest, it is a good idea to read reviews in *Personal Computing Magazine, Research in Word Processing Newsletter,* or *Software Digest.*

Why would any writer with a working typewriter want to get involved with learning a word-processing program? A typewriter will produce neatly typed documents without as much complication. Some writers cannot write without a yellow legal pad and a particular kind of pen or pencil. One doesn't have to learn commands for a program when there is a sharp pencil and a yellow pad at hand. And you can't take your computer to the beach and write while working on your tan. Computers have their limitations; they are never going to win over writers who are dependent on their present habits of composing. On the other hand, we have met few writers willing to give up their word processors once they have learned how to use them. "I could never go back," says one writer friend. "I'm hooked."

To understand the appeal of the word processor to the writer, one has to recognize the nature of the writing process. The writer does not sit down at the typewriter and produce a finished product in one scintillating swoop. Good writing is usually a matter of rewriting. Research into the writing process indicates that good writers revise continuously and at several levels while they write; poor writers do not. A possible explanation may be that good writers have sufficient confidence and control over such matters as grammar and punctuation that they are free to concentrate on revising to improve organization and precision. Poorer writers get overwhelmed by the complexity of writing; too many things must be done at once, and it is hard to sort out which are most important.

Once the writer gets a draft on paper, the real work begins. Words are deleted, others inserted. Whole paragraphs are cut out and pasted in somewhere else where they fit better into the organization. Sentences or phrases from an earlier draft are resurrected for inclusion in the new version. The writer reads over what has been written, pondering the relationships between parts. Another clean copy must be typed, and then the process of revising is repeated. And when the final clean copy is ready for the printer, the writer sees one more thing that would improve the piece. But the thought of retyping the whole paper again . . . well, let it go.

The word processor liberates the writer from many of the physical and psychological burdens that have always been part of the job. The word processor is well-suited to the tentative, exploratory nature of writing. Mistakes in typing are eliminated with a backspace or a flick of the DELETE key. Good ideas are easily inserted with the help of an INSERT key. No paper, no erasures, no white-out, no RETURN key to hit at the end of a line. The writer can type as fast as he or she is able, keeping up with the flow of ideas better than ever before. The text that appears on the monitor is more tentative than that which is typed on paper. The writer is encouraged to try out an idea, to see how it looks in print, knowing that it can be zapped with a keystroke. An electronic thesaurus can be called on when the writer is stuck for the right word, and a spelling checker will read quickly through the document, questioning the spelling of any words not in its dictionary. We know from research that good writers are always moving back and forth through the document, reading and rereading, a task made easy by the computer's capacity to "scroll" through the document with speed and ease.

If a writer can get through the frustrations of the first week of using a word processor and get comfortable with the most-used commands, he or she is likely to be hooked forever and will find it hard to return to the paper chase and its cross-outs, arrows, pasted paragraphs, and illegible marginalia. Preliminary research indicates that students find the computer a real aid in writing and that they are more likely to revise when writing on a computer than when not. On the negative side, there is also some indication that the revisions tend to be more superficial than substantive, having to do with word substitutions or corrections rather than organization and strategy. Nevertheless, if revision is so characteristic of good writing, we can only welcome any tool that encourages it and try to develop ways to channel that revision onto several levels other than mechanics.

New Programs to Help Writers

Useful as current word-processing programs are to a writer, they have been confined mainly to the transcribing and editing stages of writing. But that is changing. There is tremendous interest in how computers can be used to help teach writing. Building on the pioneering work of Major Hugh Burns at the Air Force Academy, and on the work of Bell Laboratories' *Writer's Workbench* software, other researchers have been investigating ways in which the computer can help students generate ideas, formulate a thesis, arrange the ideas in an effective order, analyze audience needs, write and edit these ideas to a disk, and then revise. The revising process is aided by computer-generated questions designed to point up problems not only in mechanics but also in style and organization. There are, for example, computer programs that ask students key questions which are to be answered in sentences. The answers are then formatted by the computer into a rough draft. Another program constructs an outline from the phrases elicited from the student by a series of questions. The new programs under development offer real promise as aids in the writing process, but they in no way will substitute for the intelligence of either the teacher or the student.

Before you rush out in search of a program that "will practically do the writing for you," remember one thing: computers can't *understand* what they "read." They can't ask searching questions about the content; they interact with the user by posing only the general questions in their program. If you provide good answers, these will help you think your way through a writing problem. The basic word for computing is still GIGO (Garbage in, garbage out).

Hugh Burns describes an amusing example of what happened with one of his early programs in invention. His student, Glenn, did not read the instructions and assumed the computer was smarter than it was. The program could not answer Glenn's content-based question, and the dialogue between Glenn and the computer program went as follows:[2]

---

[2] Hugh Burns, "Recollection of First-Generation Computer-Assisted Prewriting," in *The Computer and Composition Instruction*, ed. William Wresch (Urbana, Il.: NCTE, 1984), pp. 23–26.

NOW I NEED TO FIND OUT WHAT YOU ARE WRITING ABOUT, SO WOULD YOU
PLEASE TYPE IN YOUR SUBJECT.   I AM LOOKING FOR ONE TO THREE
WORDS.

Life

LIFE, HMMM!   WILL YOU BE AMAZED BY THE RECENT SCHOLARSHIP.   BE
SURE TO ASK THE LIBRARIAN IN THE REFERENCE AREA.
A COMMENT ABOUT PURPOSE:
DURING THIS EXPLORATION PROCESS, YOU WILL BE ASKED TO CLARIFY
THE PURPOSE OF YOUR PAPER ON LIFE.
SO NOW WOULD YOU BRIEFLY DESCRIBE WHAT THE PURPOSE OF YOUR
PAPER IS BY COMPLETING THIS STATEMENT: THE PURPOSE OF THIS
PAPER IS TO...

Refute Aristotle's theory of life

ANY MORE? (IF SO, TYPE WHATEVER IT IS. IF NOT, TYPE 'NO')

No

FINE,GLENN. YOU AND I WILL TALK AGAIN ABOUT YOUR PURPOSE.
RELAX NOW, GLENN, AND ENJOY THIS LITTLE BRAINSTORMING SESSION.
WHAT IS THE OPPOSITE OF LIFE?

Death

AHH, SHORT AND SWEET. NOW TELL ME WHY? IN OTHER WORDS,
ELABORATE A LITTLE.

Life is being, death is being no longer.

THAT'S THE IDEA, GLENN. GIVE ME SOME MORE INFO NOW.

Death is a new life and not the end.

GREAT, GLENN!
ANYTHING ELSE? (YOU CAN ADD MORE INFO, ASK A QUESTION, OR GIVE
A COMMAND--WHATEVER YOU WISH.)

What is life? (The computer, understandably, has nothing in its
program to allow it to deal with this question.)

YES, THAT SEEMS OK. YOUR TURN, GLENN.

No! You didn't answer my question!

OK   (TRY EXPLAINING A LITTLE MORE. LESS PHRASES, MORE
SENTENCES.)

QUESTION 2--ONE OF MY ALL-TIME FAVORITES COMING UP.
WHAT SPECIAL EXPERIENCES MADE YOU SELECT LIFE AS YOUR TOPIC?

Don't get off the subject pal. You tell me what life is all
about!

FINE, GLENN. WRITE SOME MORE.

You don't know what the (expletive deleted) I'm writing anyway.

IF YOU HAVE A GOOD ANSWER HERE, YOU'LL PROBABLY WRITE A DECENT
PAPER. BY"SPECIAL," I MEAN "UNIQUE," "INTERESTING," OR
"IMPORTANT."   THESE EXPERIENCES DO NOT NECESSARILY HAVE TO BE
YOURS; YOU COULD PRETEND TO BE A REPORTER.
WHAT ARE YOU THINKING NOW, GLENN?

```
I don't want to be a ...reporter.  I just need some help in
(accidentally hits RETURN key)

OUTSTANDING, GLENN!
ANYTHING ELSE?

You didn't even let me finish.  Look chump, I can play rough
too.

OKAY.
WE'RE MOVING RIGHT ALONG. HERE IS QUESTION 3.
WHAT HAS BEEN DECIDED ABOUT LIFE TO DATE?

It stinks and yours is about to end...I'm getting off this
machine.
```

Glenn's experience demonstrates the weaknesses of the first generation of prewriting software. Fortunately, it was a learning experience for Major Burns if not for Glenn.

"Virtually all of the software required for . . . comprehensive writing guidance now exists," says Peter Owens in *Popular Computing,* "but it's not yet available as a unified system.[3] The efforts to bring about that unified system are well under way. In the beginning, which is to say a few years ago, computer assistance was available mainly in the form of spell-checking programs and the electronic thesaurus. Both made you aware of choices at the editing level. More recently, computer programs have attempted to help the writer by offering some limited style analysis.

One of the most interesting efforts is a program called HOMER, which is published by Scribner's in connection with Richard Lanham's *Revising Prose* and *Revising Business Prose.*[4] The program draws attention to certain bad habits, as identified in Lanham's books, and thereby makes it possible for you to correct them. It chides you if you write a sentence that is over 40 words long. It helps you to see monotony in your sentence structure, and it counts prepositions, a sign of the "lard factor" in Lanham's view. Since it has a dictionary of vague words like *aspect* and *factors,* it can also help a writer who has gotten into bad habits of word choice recognize his or her problems. How useful is it and for whom? We tend to agree with Professor Owens: "Although designed for students, the program may be even more useful for business professionals, administrators, and bureaucrats who have been seduced by what Lanham calls 'the official style' of their disciplines."[5]

The most ambitious effort at producing a totally integrated writing assistance program was originally named WANDAH (Writing Aid and Author's

---

[3] Peter Owens, "HOMER Sharpens Prose Style," *Popular Computing,* 7 (February, 1985), 160–164.

[4] Michael E. Cohen and Richard A. Lanham, *HOMER: A Computerized Revision Program for Apple II and Apple IIE* (New York: Scribner's, 1983). The program is based on Lanham's *Revising Prose* (Scribner's, 1979) and his *Revising Business Prose* (Scribner's, 1981).

[5] Owens, p. 160.

Helper) in its developmental stage. It is now marketed commercially as *HBJ Writer.*[6]

Since this program represents a significant step in the direction of integrating all parts of the writing process, we will describe it briefly here. If your college or university computer center has a copy, you may want to give the program a try.

WANDAH was developed by the Word Processor Writing Project at UCLA, funded by a grant from the Exxon Education Foundation. Rather than evolving out of the need to transcribe text quickly and correctly, as was the case with most of the existing word-processing programs, this program was specifically designed to help university students *create* text. Based on current research into the composing process, it is intended for use by those who are generally unfamiliar with computers. As an integrated system, it aims at helping students at all stages of the process—prewriting, writing, and revising.

The system consists of three components:

1. A word processor expressly designed for on-line composing.
2. A set of four prewriting aids, allowing the student to choose whichever is most useful. These aids are based upon techniques developed by well-known researchers in the field of composition.
3. Several programs that help the student review and revise. At the level of mechanics, a limited program checks for the more obvious punctuation problems. Another checks for spelling and questions those words not in its dictionary. Still a third program looks for a hundred of the most common usage errors (like *it's* for *its*, *accept* for *except*). At the stylistic level, there is a program that helps students see certain stylistic features of their writing: overuse of prepositional phrases or abstract words, for example.

How well the system will work for the Glenns of this world remains to be seen. We know already that computers can be of considerable help in the writing and revising stages of the process. Until we have more experience with new programs, we will not know the degree to which computers can help in the all-important prewriting stage.

Until you have a PAFEO program in your computer, we remain convinced that the PAFEO process can help you more with your writing than any present computer program. In the future, we hope someone will build upon the work done in such programs as Bell Laboratories' *Writer's Workbench,* Scribner's HOMER, and Harcourt Brace Jovanovich's *HBJ Writer* to produce a program specifically designed to help those writing in business and industry.

---

[6] Developed at University of California, Los Angeles, (as WANDAH) by Morton Friedman, Earl Rand, Ruth Van Blum, Michael Cohen, Lisa Gerrard, Andrew Magpantay, Susan Cheng, Arturo Pisano, and Louis Mak. Published and distributed as *HBJ Writer* by Harcourt Brace Jovanovich (San Diego, 1986).

Index